Lecture Notes in Economics and Mathematical Systems

474

Springer
Berlin
Heidelberg
New York
Barcelona
Hong Kong
London
Milan
Paris
Singapore
Tokyo

Ulrich Schwalbe

The Core of Economies with Asymmetric Information

 Springer

Author

PD Dr. Ulrich Schwalbe
University of Mannheim
Department of Economics
D-68131 Mannheim, Germany

Library of Congress Cataloging-in-Publication Data

Schwalbe, Ulrich, 1958-
    The core of economies with asymetric information / Ulrich
Schwalbe.
      p.   cm. -- (Lecture notes in economics and mathematical
systems ; 474)
    Includes bibliographical references.
    ISBN 3-540-66028-3 (pbk. : alk. paper)
    1. Information theory in economics.   I. Title.   II. Series.
HB133.S39   1999
330'.01--dc21   99-34903
                                                                    CIP

ISSN 0075-8442
ISBN 3-540-66028-3 Springer-Verlag Berlin Heidelberg New York

© Springer-Verlag Berlin Heidelberg 1999
Printed in Germany

Typesetting: Camera ready by author
SPIN: 10699869   42/3143-543210 - Printed on acid-free paper

Acknowledgements

I am very much indebted to Volker Böhm and Jürgen von Hagen for their encouragement, their help, and many stimulating discussions during the time the conceptions presented here were developed. They have read an earlier version of the manuscript and provided many detailed comments and valuable suggestions. I also want to express my deep gratitude to Tone Dieckmann with whom I have discussed my ideas at length many times. Without her constant support and her advice I could never have completed the work. Furthermore, I am very grateful to Hartmut Stein who helped to derive many of the mathematical results. I also wish to thank Siegfried Berninghaus, Tatsuro Ichiishi, Jörg Naeve, Walter Trockel, and Nicholas Yannelis who made many helpful comments and suggestions. Finally, my thanks go to Martina Bihn and Werner Müller from Springer Verlag for managing the project in a very efficient and friendly way.

March 1999

Ulrich Schwalbe

Contents

Chapter 1

Introduction – Information in Models of General Equilibrium

Few would disagree that information is a central issue in economic analysis. Information matters not only on the level of the individual decision maker, but also with respect to the functioning of markets and, what is more, entire economies. While studies of the effects of information on an economy can be traced back to von Hayek (1937),[1] it is only in the last 25 years that an independent area of research has been established. Under the heading of *economics of information*, this area takes on a central position in microeconomic theory. However, up until today, no single generally accepted definition of *economics of information* has been established. The main reason is that the term is applied to various different fields. While some authors view information economics as a certain branch of economic theory, such as e. g. environmental or health economics,[2] others argue that informational problems play a crucial role in *all* aspects of economics,

---

[1] See von Hayek (1937)

[2] 'The economics of imperfect information is therefore an integral part of the theory of industrial organization.' Phlips (1988), p.4.

and should therefore constitute a part of every area of economic theory.[3]

The spectrum covered by information economics today ranges from Stigler's search theory[4] to industrial economics, including oligopoly theory, innovation, as well as research and development.[5] However, the area information economics is most closely connected with is the theory of optimal contracts, mainly analyzed in principal–agent models.[6] Contract theory deals primarily with the question of how optimal arrangements (contracts) for the purchase and sale of commodities and services between two or more agents should be structured. In these models, it is often assumed that the parties to the contract are informed differently or asymmetrically about relevant variables (e.g. the health of one party in the case of insurance contracts, or the effort in relation to employment contracts). As a result of this asymmetric information, phenomena such as moral hazard, adverse selection, signaling, and screening may arise. Frequently, results from contract theory are referred to when making statements about the effects of asymmetric information on an economy. Models of this kind are often used to explain phenomena such as fixed wages or unemployment, among others.[7]

However, such conclusions must be treated with caution for two reasons. In the first place, in these models, a contract (explicit or implicit) is determined by the solution of an optimization problem. However, whether a phenomenon such as unemployment can be considered a result of contracts constructed in this way

---

[3]'I wanted to suggest that informational considerations were, in fact, central to the analysis of a wide variety of phenomena, that they constituted a central part of the Foundations of Economic Analysis.' Stiglitz (1984), p.21.

[4]See Stigler (1961).

[5]See for example Reinganum (1983) and Dasgupta and Stiglitz (1980).

[6]An overview of the theory of contracts theory can be found in e.g. Hart and Holmström (1988) or Salanié (1996).

[7]'Thus, the empirically observed phenomenon of sticky real wages and layoffs may in fact be the outcome of jointly optimizing behaviour on the part of workers and firms who enter into long-term, unwritten agreements.' Taylor (1988), p.139.

is, in my opinion, doubtful. It is rather a question of equilibrium or rather disequilibrium. Secondly, the vast majority of studies that assume asymmetric information employ a partial equilibrium framework – in many cases these models consider only two agents. It is therefore at least doubtful whether conclusions about the effects of asymmetric information on a whole economy can be drawn on the basis of these partial equilibrium approaches. In addition, the majority of the contract theoretical approaches cannot be applied directly to a model of an economy, i.e. to a general equilibrium model, mainly because of their specific formulation. This is one of the reasons why – considering the large number of partial equilibrium models in which asymmetric information is analyzed – only a very modest number of papers deal with the effects of asymmetric information for an economy as a whole.

The models of general equilibrium with asymmetric information published so far can roughly be divided into three categories. The criteria for such a division are the description of the information and the equilibrium concept.

Two aspects of information are particularly emphasized in economic theory: on the one hand, it is stressed that information shares certain properties of a commodity in the usual sense and therefore – similarly to other commodities – can be exchanged or bought or sold at the respective markets for information.[8] On the other hand, it is emphasized that the information available to an agent should be regarded as a characteristic of this agent. If this is the case, the information of an agent describes – intuitively speaking – his knowledge of the existence of exogenous facts, so-called states of nature. With this type of modeling, an agent's information is viewed in the same way as his other characteristics, i.e. his initial endowment of commodities and his preference relation.

With regard to the equilibrium concept used in the literature, two approaches can be distinguished: models that are based on the

---

[8]'Thus, information is not merely a good that is desired and acquired but is to some extent a commodity like others whose markets we study.' Arrow (1973), p.142.

Walrasian equilibrium is employed, and models which rely on the core of an economy. These different approaches to the problem of asymmetric information in a general equilibrium framework will be illustrated by several representative works, and the respective advantages and disadvantages of the modeling will be discussed.

1.1   Information as a Commodity

Among the major works that treat information as a commodity are the models of Beth Allen (1986, 1990). She considers a set of possible states of nature, where a state of nature is a complete description of all economically relevant variables. An information is defined by a sigma algebra on this set of states of nature, where different information is described by different sigma algebras. An agents's information constrains his consumption plans, i.e. elements of the set of physical commodities, because only those consumption plans are admissible which are compatible with his information, i.e. which are measurable with respect to his sigma algebra. If an agent cannot differentiate between two states of nature, then he cannot consume different amounts in these states.

Allen regards different information as differentiated commodities, which are further assumed to be indivisible.[9] This assumption avoids the problem of the quantification of information if it is considered as a commodity, as pointed out by Arrow (1973).[10] Besides his initial endowment of physical commodities, an agent also possesses an initial endowment of information. This initial endowment of information consists of a finite number of sigma algebras. It is assumed that the information itself generates no

---

[9] 'Yet another problem ... that arises when one attempts to analyze information as an economic commodity is that information structures are inherently indivisible and useful only in integer amounts.' Allen (1990), p.270.

[10] 'In particular, the well–known Shannon measure which has been so useful in communications engineering is not in general appropriate for economic analysis because it gives no weight to the value of information.' Arrow (1973), p.138.

utility, i.e. the preferences of an agent are defined only for physical commodities or state contingent commodities. The consumer is interested in information only in so far as additional or better information will allow him the choice of a better consumption plan, because a finer sigma algebra weakens the required measurability condition on the consumption plans. An agent's preferences with respect to information are therefore derived preferences, comprising the prices of the physical commodities as well as the value of the initial endowment. Furthermore, it is assumed that demand will be satiated with one unit of information, because the agent's possible courses of action cannot be increased by two or more units of the same information.

In order to prove the existence of a Walras equilibrium, Allen considers an economy in distribution form. In other words, she does not consider an Arrow–Debreu economy in the traditional sense.[11] The equilibrium concept employed is an equilibrium distribution over the space of characteristics of the agents as well as the prices of information and the prices of the physical commodities. Allen shows that such an equilibrium distribution exists in her model. This equilibrium is characterized as follows: All markets for information are cleared and the average excess demand for each physical commodity aggregated over the states of nature is equal to zero.

However, a number of problems arises from this formulation of information: If information is treated as a commodity, the question arises how the transfer of ownership of a unit of information should be perceived when interpreting the purchase or sale of information. Does the person who is selling the information lose this information, or is he able to sell it and at the same time keep an identical copy of it? This problem has been pointed out, among others, by Hirshleifer and Riley (1992): 'It would be possible to transmit purely private information only if a technology were to emerge whereby *forgetting* could be reliably effectuated. Then, for a price, I might give you some news while arranging to

---

[11] On the theory of economies in distribution form see Hildenbrand (1974), p.124-127 or Ellickson (1993), p.127-132.

forget it myself!'[12] Allen's statements concerning this point are inconsistent: on the one hand, she admits that a seller remembers the information he sells, but on the other hand she rules out the possibility that a consumer is capable of making an unlimited number of copies of his information, because this would reduce the equilibrium price for information to zero.[13] A further problem is associated with the purchase or sale of a piece of information: if it is assumed that a piece of information can be traded, then the buyer of the information cannot be sure that he is receiving the *correct* information, because he is not in a position to check its correctness. Allen rules out this possibility by assuming the markets for information to be sufficiently 'thick', so that the agents do not behave strategically.[14] Because she continues to model information as a commodity by a sigma algebra, she treats it as if it were detached from a material information carrier. Yet this modeling does not correspond to Allen's interpretation concerning information as a commodity.[15] Moreover, as Hirshleifer and Riley point out, a piece of information itself cannot be purchased: 'Since you can never know in advance what you will be learning, you can never purchase a *message* but only a message service – a set of possible alternative messages.'[16]

---

[12]Hirshleifer and Riley (1992), p.169.

[13]'... traders can remember some of their information after it has been sold to another agent ...' (Allen (1986), p.11. '...making many copies or using the information and then passing it along to someone else is precluded.' Allen (1986), p.3.

[14]'This alleviates the incentives problems associated with moral hazard and lies, as well as the possibilities for strategically withholding information from others.' Allen (1988), p.4.

[15]'The reader should imagine that the information here takes the form of a written report or data on a computer tape. ...Some other examples of the information considered in my paper include magazines and newspapers, various financial newsletters, scientific instruments ....' Allen (1988), p.2.

[16]Hirshleifer and Riley (1992), p.168. The concept *message* used by Hirshleifer and Riley corresponds to an information in the framework considered here.

Due to the problems mentioned, the approach of integrating information as a commodity with special properties in a model of general equilibrium appears in some respects unsatisfactory.

1.2 Information as a Characteristic of an Agent

Apart from the approach outlined above, there are several models of general equilibrium in which an agent's information is interpreted as one of his characteristics, comparable for instance with his preference relation or his initial endowment of commodities. Radner's work (1968) is particularly relevant in this context. Here, as in Allen's model, it is presumed that uncertainty can be modeled by a number of possible states of nature. The information of an agent, i.e. of a consumer or a firm, is defined by a sigma algebra or a partition of the set of states. Extending the model of an economy with uncertainty, as it was first introduced by Debreu in Chapter 7 of his 'Theory of Value', Radner allows the agents to be asymmetrically informed about the states of nature. This asymmetric or differential information is introduced by assigning different sigma algebras or partitions of the set of states to the agents. It is assumed that, if an agent is incompletely informed about the states of nature, he can only carry out such consumption or production plans that are compatible with his information. An agent cannot make a consumption or production plan contingent on events without being informed of their occurrence.

Radner then analyzes the existence of a Walrasian equilibrium and shows – under the usual regularity conditions on the preferences, consumption and technology sets – that such an equilibrium exists.

In contrast to Allen's approach, Radner's model does not encounter such serious conceptional problems with respect to the modeling and interpretation of information. His model is an extension of the original model of Arrow–Debreu: if all agents have the same information at their disposal, then the usual model of an economy with uncertainty will result. However, an equilibrium in this extension of the Arrow–Debreu model is often characterized by the fact that only those net trades are compatible with an equilibrium which are independent of the states of nature. The reason for this somewhat unsatisfactory result can be explained as follows:

Firstly, net trades can only be carried out contingent on events observed by *all* agents together.[17] However, as Radner himself argues, this is a considerable restriction: 'In practice, since agents make contracts with other agents, and not with an abstract "market", delivery will be contingent upon information that is common to the two agents in question ...'[18] So it is not the information of all agents in the economy that is relevant to the transaction, it is only the information which is available to those involved in the respective transactions. Secondly, Radner's model does not take into account the possibility of an exchange of information between the agents.[19] The third reason for this result can be explained as follows: only a certain type of arrangements for carrying out transactions is considered, namely contracts with linear prices. Radner is aware of this limitation, because he writes: 'In real markets, various types of contract, not formally admitted in the extended Arrow–Debreu model, have the effect of partially circumventing these restrictions.'[20]

---

[17]'... the net trade between any group of agents and the group of all other agents in the economy can at most depend upon information that is common to both groups of agents.' Radner (1968), p.50

[18]Radner (1968), p.50.

[19]Radner provides an example where he considers the possibility that an agent can use resources to change his information. However, this generally leads to the occurrence of nonconvexities in the consumption and technology sets, so that the usual conditions for the existence of a Walrasian equilibrium are, in general, not fulfilled.

[20]Radner (1982), p.948

In the following years, Radner's model was extended in several directions, whereby the question mentioned by Radner as to alternative contract forms became less significant. The extensions served, in the first place, to integrate into the model other phenomena characterizing a market economy, such as money, stock markets and – in an intertemporal context – active markets at any point in time. Particularly due to this last point, in the extensions of Radner's model it is assumed that at any point in time the market system is incomplete, so that the agents are uncertain not only about the states of nature, but also about future prices. Therefore, questions concerning the formation of agents' expectations play a central role in models of this kind.[21] In this context, an extensive range of literature has been published in recent years, which can be classified under the title of 'Equilibria with rational expectations'.[22] These models also raised the question was also raised whether agents can draw any conclusions about the non-price information of other agents by observing the equilibrium prices.[23]

The third group of models also describes an agent's information as a partition of a set of possible states of nature, but it uses the core of an economy as the equilibrium concept. These works, the majority of which originate from the 1990s, go back to a paper by Wilson from 1978. In this article, Wilson mainly investigates the question which allocations can be characterized as efficient when the available information is asymmetricly distributed. In particular, he considers the possibility that agents are able to exchange information by using a communication system. The approaches he has developed since then have extended and modified his original model. These models are described systematically in the first chapter of this study and will not be discussed in more detail at this point.[24]

---

[21] See e.g. Radner (1972).

[22] For a survey of this area of research, cf. Jordan and Radner (1982).

[23] See e.g. Allen (1981) or Radner (1986).

[24] Besides the different models mentioned, there is extensive literature in the field of welfare economics, in which the question of the implementation of a social choice function is investigated. The assumption is frequently made

In comparison with the Walras equilibrium, the concept of the core of an economy has a number of advantages with regard to the modeling of asymmetric information. First, an analogy to contract theory mentioned above can be drawn as follows: The cooperative approach in these models allows the agents to form coalitions, i.e. binding agreements. If such a binding agreement is interpreted as a contract, the cooperative approach allows for the transfer of some concepts from contract theory, which mainly is a partial equilibrium approach, to a model of general equilibrium. Furthermore, models based on the core allow for the possibility that agents can exchange or aggregate information. This establishes a link to models in which information is considered a commodity. On the basis of these advantages, the following enquiry will refer to the equilibrium concept of the core for the analysis of an economy with asymmetric information.

1.3   Organization of the Analysis

This study is organized as follows: the first chapter provides a survey of the literature on the core of exchange economies with asymmetric information. The available models will be systematically introduced and discussed. The second chapter investigates the core of an exchange economy with asymmetric information. The central result of this chapter is that, irrespective of whether or not, and in what way, the agents are able to exchange information amongst each other, the core of such an economy is nonempty. In the third chapter, the model is extended in order to allow for production. This is done by assigning a technology set to each coalition. This model of a coalition production economy with asymmetric information is an extension of Böhm's model

---

that the agents or planners are informed asymmetrically about relevant parameters (e.g. preferences). As the main concern of the welfare theoretic contributions is not the analysis of a given economy with asymmetric information, but the construction of incentive compatible mechanisms, this literature will not be dealt with here. For an overview of these models, cf. Moore (1990) and Palfrey (1990).

(1974a). As in his model, a certain property of the technology sets – their balancedness – plays a crucial role for the existence of core allocations. The impact of asymmetric information on such a coalition production economy is illustrated by a simple example. In addition, the case of costly coalition formation is costly is dealt with, and it is shown that the result that the core of the economy is nonempty still holds. The study closes with a discussion of the results and several extensions will be suggested.

Chapter 2

Survey of the Literature

This chapter provides a systematic survey on the literature about the core of exchange economies with asymmetric information. The assumptions of the different models will be ordered according to certain aspects. The different approaches to model e.g. the information of the agents, the exchange of information between coalition members, and feasible allocations, will be compared, and the impact of the various assumptions on the results of the models will be discussed.

2.1   The Information of the Agents

As mentioned in the introduction, the analysis of economies with asymmetric information has employed two different ways to model information. One of them views information as a special commodity which can be exchanged or sold and purchased on markets (Allen (1986), (1990)). The other formulation describes information as a characteristic of an agent, comparable to his preferences or his initial endowment of commodities. All models of the core of an exchange economy with asymmetric information published up until now – beginning with Wilson's model in 1978 to the recent papers by Volij (1997), and Lee (1998) – have used the second type of model.

Modeling information in this way can be traced back to an article by Radner (1968), in which asymmetric information was introduced into a model of general equilibrium for the first time. Radner assumes a measurable space consisting of a set of possible *states of nature* and a number of measurable subsets of this set of states, the measurable. *events* A state of nature describes all variables regarded as parameters for a given economic model. These may include the weather, the occurrence of a natural disaster, or, at an individual level, the state of an agent's health. An agent's information about the states of nature can be described by a sigma algebra or, for the case of finitely many states of nature, by a partition of the set of states.

The agent's information thus describes which events this agent can discern. If an element of the sigma algebra comprises the partition of several states of nature, the agent will not be able to differentiate between these states. This concept of information is often interpreted in the literature in the following way: each agent has access only to a vague measuring instrument which he uses to observe the world. The signals he receives from his instrument may therefore be rather coarse. However, only these measurement results allow the agent to draw conclusions about the states of nature. Therefore, his information about the states of nature will be imperfect.

While in earlier literature on the core of an economy with asymmetric information a finite set of states of nature was assumed (e.g. Wilson (1978), Srivastava (1984a, 1984b)), more recent studies (Allen (1991a, 1991b, 1992), Yannelis (1991), Koutsougeras and Yannelis (1994, 1995), Vohra (1997), Volij (1997), Lee (1998)) allow for an infinite number of states of nature.

2.2 Commodities, Consumption Sets, and Initial Endowments

All studies of the core of an economy with asymmetric information retain the usual assumption that there is a finite number of physical commodities. The physical commodity space is modeled as a Euclidean space, where the number of physical commodities available in the economy determines the dimension of the physical commodity space. However, the presence of uncertainty, as described by the set of states of nature, implies that it is not the physical commodity space, but the space of state–dependent or *contingent* commodities that is the relevant concept for analysis. Depending on whether the number of states of nature is assumed to be finite or infinite, it follows that the commodity space has finite or infinite dimension. For the analysis of a model with an infinite–dimensional commodity space, such as Allen (1991a, 1991b), Yannelis (1991) and Koutsougeras and Yannelis (1994), more advanced methods from measure theory or functional analysis, or the theory of Banach lattices must be referred to. However, it is not clear whether the economic content of these studies would increase with this generalization – allowing for an infinite dimensional commodity space has not considerably changed the economic results of the models.

As far as the consumption set is concerned, i.e. the subset of the commodity space from which an agent can choose his consumption plans, two alternative assumptions are made in the literature: while most studies assume that the consumption set is a given subset of the commodity space, some models (e.g. Wilson (1978)) allow the set of consumption plans which can be chosen by an agent to vary with the states of nature, i.e. the consumption set is modeled as a correspondence from the set of states of nature to the commodity space. In both types of formulation, however, a consumption set is assumed to be closed and bounded from below.

Initial endowments of commodities are defined as nonnegative elements of the commodity space or, equivalently, as functions from the set of states of nature to the consumption set. Furthermore, it is mentioned (e.g. Wilson (1978)) that the agents possess an initial endowment of information.[1] However, the term initial endowment in the context of information is confusing, because information is on a different conceptional level as the initial endowment of commodities, as will be discussed in more detail in section 2.4. In all models, information is considered only as a constraint which restricts the consumption plans an agent may choose.

2.3   Preferences

As with the formulation of the commodity space, the consumption set, and the initial endowment, there is a large degree of consensus in the literature with respect to the modeling of the agents' preferences. Almost all studies assume that the agents' preferences can be described by an expected utility function, i.e. the concept of probability is explicitly allowed for in the analysis. This is not the case in, for example, Chapter 7 of Debreu's 'Theory of Value' or in Radner's model. It is assumed that each agent has a probability distribution on the set of possible states of nature. Some models (e.g. Allen (1991a, 1994a)) allow for the assumption that this probability distribution can vary across the individual agents (i.e. so–called 'subjective' probability distributions are possible). Furthermore, in some papers (e.g. Yannelis (1991)) it is assumed that an agent's preferences may depend on the state of nature, i.e. state–dependent utility functions are taken into consideration.

However, such a description of preferences based on probability theoretical is not necessary: as Debreu explains in Chapter 7 of his Theory of Value, preferences under uncertainty can be formulated without referring to a probabilistic specification. If preferences are defined for contingent commodities, the preference relation itself reflects the personal estimation of the probability that a certain event occurs. The main advantage of this assumption is that

---

[1] '... an economy in which agents differ in their endowments of information.' Wilson (1978), p.807.

the analysis is not limited to the set of economies in which the agents' preferences can be described by an expected utility function. While this formulation of preferences is common to general equilibrium models with symmetric information, almost all works on asymmetric information rely on the concept of expected utility, with the only exception of Srivastava (1984a, 1984b).

The objective of an agent is to maximize his utility by selecting an optimal consumption plan. The literature differentiates whether the agent chooses his consumption plan *before* or *after* an event has occurred. Thus, there are models with *ex ante* utility maximization, such as Allen (1991a, 1991b, 1992, 1994a) or Koutsougeras and Yannelis (1994), and with a utility maximization *ex post*, of which the studies of Wilson (1978) and Yannelis (1991) are examples. While in the ex ante models, the agents maximize their expected utility, in the ex post models, the objective function is the *conditional* utility function.[2]

This difference in the objective function of the agents is referred to in the literature (e.g. Allen) as the reason for the somewhat differing results.[3] However, Koutsougeras and Yannelis have established that the assumption of an ex ante or ex post utility maximization is not decisive for the results.[4] As will become clear later on (sections 2.6 and 2.10), the different results of the various models arise mainly from different definitions of feasible allocations.

---

[2] *Ex post* is sometimes called *interim* utility maximization

[3] 'The primary difference is that in Wilson's (1978) model, blocking occurs ex post whereas I use an ex ante core concept.' Allen (1991a), p.25.

[4] 'Note that all the results of this paper remain valid if we choose the alternative conditional expected utility for formulation.' Koutsougeras and Yannelis (1994), p.6.

2.4   The Impact of Information on Consumption Plans

All models of the core of an economy with asymmetric information share the same basic assumption regarding the impact of information on an agent's consumption decisions: if an agent is incompletely informed about the states of nature, this imperfect information restricts his choice of consumption plans. However, the various models use different specifications of this constraint.

The assumption that imperfect information implies a restriction of the admissible consumption plans has been introduced in Radner's model (1978). He assumes that an agent can only choose consumption plans that are consistent with his information about the states of nature. In other words, admissible consumption plans have to be measurable with respect to his information.[5] This assumption is justified by the fact that an agent cannot choose different consumption plans contingent on states which he cannot discern. This assumption of measurability is required not only with respect to the consumption plans, but also the initial endowment. Otherwise, it is claimed, an agent could obtain additional information simply by observing his initial endowment.[6] If he possessed different initial endowments in two states, he would be able differentiate between the two states. However, it is not explained what the 'observation of an initial endowment' means in a formal sense. The requirement of measurability of the initial endowments as well as the consumption plans obviously implies that the net trades of an agent are also compatible with his information.

---

[5] In the case of finitely many states of the world, the assumption of measurability of a consumption plan is equivalent to their being constant on an element of his information, a partition of the set of states.

[6] 'If we suppose, as we should, that the information structure takes into account all sources of information, we must naturally assume $\omega_i \in A(S_i)$; otherwise, the initial endowment would give the consumer additional information.' Guesnerie and de Montbrial (1974) p.61. Here ω_i denotes the initial endowment of agent i and $A(S_i)$ the set of all functions which are constant on an element of an information S_i of agent i.

In the literature on the core of an economy with asymmetric information, several assumptions are made with respect to the measurability of initial endowments, consumption plans and net trades. Thus Wilson asserts that, while the consumption plans must be consistent with the agent's information, no conditions are stipulated as to the measurability of the initial endowment. Consequently, the net trades that an agent can carry out need not be consistent with his information. This model thus accepts that an agent can also carry out net trades in states of nature between which he cannot differentiate, and is therefore able to determine his trading on events of whose occurrence he is not informed.

This inconsistency leads Srivastava (1984a) to criticize Wilson's assumption who postulates that, not only the consumption plans must be measurable, but also the initial endowment (and thus implicitly also the net trades) have to be compatible with an agent's information. He even goes one step further, demanding not only the consistency of the initial endowment with the information, but also supposing that an agent obtains his information about the states of nature exclusively by making inferences from his initial endowment. He can differentiate between these states only when his initial endowment differs between states of nature. An agent's information in Srivastava's model is thus not given exogenously given. Instead, it is a derived concept. For example, if this condition is satisfied, an initial endowment which gives him the same amount of physical commodities in each state of nature would not lead to any information at all for an agent. Such a claim is, however, excessively restrictive, because it is not clear why an agent should not possess precise information about the states of nature, even if he possesses the same initial endowment of physical commodities in each state.

Yannelis (1991) and Koutsougeras and Yannelis (1994) make a somewhat weaker claim, identical to Radner's assumption. In this case, it is not taken for granted that the initial endowment is an agent's only source of information. The only requirement is that his initial endowment be compatible with his information. In these models, an identical initial endowment is certainly compatible with complete information in all states, enabling the agent to differentiate between all states of nature. The least restrictive assumption on the impact of information on the decisions of an agent is made by Allen (1991a, 1991b, 1994a). In her models, information is determined neither by the measurability of the initial endowment, nor by the consumption plans. She only requires that the agents' net trades be compatible with their information. In other words, there is no relationship between the information of an agent and his initial endowment. Information has an impact on his consumption plans only in so far as it limits his net trades. The information is therefore relevant to what an agent can do, but irrelevant to what he has got as initial endowment. Above all, the measurability of the net transactions is, as will become clear later (see section 2.10), a central requirement for proving the existence of core allocations. The compatibility of the initial endowments or consumption plans with the information is not important here.

2.5   Information of the Agents as Coalition Members

When analyzing the core of an economy, the possible coalitions and the allocations that they can enforce are of primary importance. However, if the agents are asymmetrically informed about the states of nature, the question arises which information the agents can use as members of a coalition in order to reallocate their initial endowments of commodities. It might be supposed that, when agents form a coalition, they have the possibility to communicate with each other and to exchange their information. This idea was first modeled by Wilson, using the concept of a *communication system*. In this model, he assumes that communication can only improve the information of an agent but will never impair it. Stated otherwise, the number of events between which an agent can differentiate will not diminish as a result of communication. Wilson considers different communication systems in his analysis. He defines the *perfect* communication system which pools the information of all agents, i.e. leads to the finest information for each agent. Using a perfect communication system, the agents can share their information completely and reliably with the other coalition members. Thus, an agent cannot obtain any better information than that which he receives through perfect communication with all other agents. If, however, a communication system offers no possibilities for the exchange of information, Wilson refers to a null communication system. In this case, the information available to a coalition member will not change. All other communication systems he considers are intermediate cases between the perfect and the null communication system. Srivastava (1984a,1984b) makes the same assumptions with respect to the exchange of information between coalition members.

These assumptions rule out the case that the agents as coalition members have to employ information which is inferior to the information they were initially given. This case might arise, for example, when the agents can carry out reallocations of their endowments only contingent on events *jointly* observed by all coalition members. This assumption excludes the possibility that the agents also receive information from other sources than from a pooling of the information of all agents.

However, in Wilson's model a communication system represents only an upper bound for the exchange of information. Even if a coalition has a perfect communication system at its disposal, it is not forced to use this system. It can also renounce an exchange of information. Therefore, the coalition has a choice between a number of communication systems. However, Wilson makes no statement as to how and according to what criteria a coalition decides on the communication system to be employed.

Wilson's idea to model the exchange of information between coalition members by a communication system has since been employed in some form or other by all subsequent models. For example, Yannelis (1991) makes the very restrictive assumption that the agents are unable to exchange information among each other. That is, he assumes the null communication system. Yannelis gives the following argument for ruling out the exchange of information: 'Finally, it is worth pointing out that a core notion which allows for complete exchange of information among agents in each coalition may not be an appropriate concept since in most applications, agents do not have an incentive to reveal their own private information (think of situations of moral hazard or adverse selection).'[7]

So far, the most satisfactory formal modeling of the information of agents as coalition members was developed by Allen (1991a, 1991b, 1994a). She introduced the concept of an *information sharing rule* (henceforth referred to as *information rule*), which allows a description of all types of information exchange in a compact form. Formally, an information rule is defined by a mapping which assigns, for every coalition, an information to every member. This information will generally depend on the information of all coalition members. With the concept of information rules, all cases discussed in the literature can be considered. For example, the perfect communication system referred to by Wilson can be modeled by an information rule which assigns to every coalition member the 'pooled' information, i.e. the coarsest common refinement of the information of all coalition members. The null communication system is illustrated by an information rule which leaves the information of every coalition member unchanged, in other words the identity mapping. Furthermore, the concept of an information rule also allows for the description and analysis of situations which until then had not been considered in the literature. For instance, an information rule can be used to investigate the case where the agents, although they possibly possess 'good' information, can use no information at all as coalition members.

Describing the information available to an agent as a coalition member by an information rule is up to now the most general formulation of the problem of information aggregation. Furthermore, it offers a simple way of investigating what consequences different forms of information exchange – characterized by differ-

---

[7] Yannelis (1991), p.188

ent information rules with perhaps different properties – have on the existence and properties of core allocations.

However, to model the information of an agent as a coalition member is not entirely satisfactory. In a recent paper, Volij (1997) has attempted to endogenize the information transmission between coalition members. Starting from Wilson's (1978) paper, Volij allows the members of a coalition to repeatedly express their willingness to trade. Since Volij assumes finitely many states of nature, this process necessarily converges within a finite number of rounds in this dialogue between the members of the coalition. Thus, it becomes common knowledge among the members of a coalition if there is an allocation which would make everybody in the coalition better off and this information is taken as the information the coalition members can employ to block a given allocation.

A different approach to endogenize the information of an agent as coalition member has been proposed by Koutsougeras and Yannelis (1995). They assume that agents start making transactions using their private information. However, they are able to observe the resulting allocation and thereby to make inferences about the information of other agents, i.e. to learn. This new information gives rise to a further round of trading and a different allocation which generates new information and so on. Koutsougeras and Yannelis consider the limit of this process. If there are only finitely many states of the world, the process will converge to perfect information for every agent. However, if the number of states is infinite, the limit may differ from perfect information as the learning process could get stuck before the perfect information has been reached.

2.6   Feasible Allocations

The theory of optimal contracts has pointed out early that in the presence of information problems, the definition of the term 'feasible allocations' is of major importance. In this context, a distinction has to be made between a first–best solution, the result that would arise if there were no information problems, and second–best solution, which arises when agents are asymmetrically informed. Due to asymmetric information, the first–best solution is frequently infeasible as emphasized in the literature.[8] In other words, the set of feasible allocations is therefore, among other things, determined by the information of the agents.

It might be supposed that this not only applies to partial equilibrium models, but also holds true in a general equilibrium framework. A feasible allocation has, of course, to be at least physically feasible, i.e. no more commodities must be distributed than are available. Furthermore, the information available in the economy has an important impact if ex ante models are considered. Intuitively, one might suspect that, for two economies which only differ with respect to the agents' information, feasible allocations and information are related in the following way: in an economy in which the agents possess 'more' information, the set of feasible allocations is larger. Furthermore, if agents are able to exchange information, then the concept of feasibility should take this into account. Here too, one might suspect that, in an economy in which information exchange is possible, more allocations are feasible than in an otherwise identical economy in which information cannot be exchanged. Therefore it might happen that in an economy in which the agents possess only a limited amount information, but in which information exchange is possible, the set of feasible allocations is larger than in an otherwise identical economy with 'well–informed' agents in which there is no exchange of information.

The impact of information and the exchange of information on the feasibility of allocations has been discussed in the literature only in passing. The earliest studies of the core 3of an economy with asymmetric information by Wilson (1978) and Srivastava

---

[8]In their textbook Milgrom and Roberts (1992) formulate thus: 'With private information, the full–information efficient solution may no longer be feasible.' Milgrom and Roberts (1992), p.141.

(1984a, 1984b) make no further statement about the feasibility of allocations – apart from the physical feasibility. This holds also true for the recent work by Volij (1997). The reason is that they consider ex post utility maximization, i.e they don't suppose that agents have to sign contracts for all contingencies.

Yannelis (1991) assumes that the feasible allocations have to be compatible with the information of the individual agents. As he does not consider the possibility of an exchange of information between agents, it is of no consequence for his definition of feasibility. In addition, Koutsougeras and Yannelis (1994) have proposed two alternative definitions of feasibility.[9] However, these are not derived from the possibilities of information exchange which exist in the economy, but have been introduced ad hoc.

The only studies in which both the information available in the economy and the possibilities of information exchange enter into the concept of feasibility are by Allen (1991a, 1991b, 1994a). She requires that an allocation has to be generated by net trades which are compatible with the information the agents can employ as members of the *grand coalition*. Obviously, this information, and hence the set of feasible allocations, depends on both the information of the individual agents and the information rule. The special role of the grand coalition for the definition of feasible allocations stems from the theory of cooperative games which usually assumes that games are superadditive. In these games, the grand coalition can achieve at least as much as what smaller coalitions can attain. However, in games where information is important, a problem arises with information rules where the information an agent can employ is getting coarser the larger the coalition becomes (think of the case in which reallocations can only be carried out contingent on events observed by all coalition members). This problem can be illustrated by the following example: In an economy where all agents possess perfect information, there are no information–related constraints on the set of feasible allocations or the respective net trades. If, however, the information of only

---

[9]These definitions are introduced in connection with two concepts of the core, the strong coarse core and the weak fine core. Cf. section 2.9.

one agent is changed, so that he is unable to discriminate between the states of nature, the set of feasible allocations is drastically reduced. The only allocations which are now feasible are those which can be generated by net trades which are constant *for every agent*. In this case, by assuming the grand coalition to be decisive for feasibility, insufficient emphasis is placed on the information available in the economy. As discussed in more detail in section 2.10, the different definitions of feasibility of allocations – beside the various assumptions about the measurability of initial endowments, consumption plans and net trades – play a crucial role with regard to the sometimes contradictory results of the models.

2.7   The Blocking of an Allocation by a Coalition

To analyze the core of an economy, the blocking of an allocation by a coalition has to be defined. In general, a coalition will always block a given allocation when its resources enable it to enforce another allocation which, for strictly monotonic preferences, will give all members of the coalition a higher utility. Which allocations a coalition can enforce depends, of course, on the initial endowments of its members, as well as their information. As in the case of feasibility, allocations enforceable by a coalition have to be at least physically feasible for the coalition. Therefore, no more commodities may be distributed than the coalition members possess as initial endowments.

The literature makes a number of different assumptions with respect to the impact of information and the possibilities of information exchange on the blocking of an allocation. These differences arise, on the one hand, from different requirements concerning the measurability of the initial endowments, allocations and net trades, and on the other hand, from the various assumptions about the possibilities of information exchange between coalition members. For this reason, Wilson postulates that an allocation that can be used for blocking has to be compatible with the information available to the members using a communication system. A perfect communication system implies the measurability with respect to the 'pooled' information, i.e. the coarsest common refinement of the information of the coalition members. Since Wilson makes no assumptions with respect to the measurability of the initial endowments, feasible allocations can be achieved through net trades which are not necessarily compatible with the information of the coalition members. This assumption is unsatisfactory, as stated in section 2.4. Since Srivastava (1984a, 1984b), as well as Yannelis (1991) and Koutsougeras and Yannelis (1994), assume not only the measurability of the allocations, but also of the initial endowments and hence the net trades with respect to the information available to the agents as members of a coalition, this problem does not arise here: feasible allocations must be compatible with this information and, furthermore, have to be feasible through net trades which are compatible with this information. The same applies to the models of Allen (1991a, 1991b, 1994a) who, however, requires only the compatibility of the net trades with the information of the coalition members – but not the compatibility of the initial endowments and the allocations. All allocations attained through net trades which are measurable according to this information can be used to block a given allocation.

The sets of allocations enforceable by a coalition vary according to the assumptions concerning the exchange of information between the coalition members: the more difficult it is to exchange information, the smaller the set of feasible allocations. If, for example, there is no exchange of information between the coalition members at all, then a coalition can use only those allocations which are measurable with respect to events observable by all coalition members for blocking. Thus, the measurability requirements in this case are very strict.

In addition, the definitions of blocking vary with respect to the assumptions about utility maximization, if the blocking of an allocation is viewed *ex ante* (cf. Allen (1991a, 1991b, 1992, 1994a),

Koutsougeras and Yannelis (1994)) or *ex post* (cf. Wilson (1978),
Yannelis (1991)). When an allocation is blocked ex post, it is assumed that a coalition can block a given allocation only after the
event has occurred. A blocking can be carried out only contingent
on the event having taken place. As in these models the agents
are assumed to be characterized by expected utility functions, an
allocation used to block a given allocation must guarantee the
coalition members a higher *conditional* expected utility. On the
other hand, blocking ex ante means that a coalition must block
an allocation with the help of another allocation *before* an event
takes place. This corresponds to the assumption usually made in
the theory of general equilibrium under uncertainty: agents sign
contracts about the supply and receiving of contingent commodities, and effect payments before an event has taken place. When
an event occurs, the transactions specified in the contracts are
carried out.[10] In order to block an allocation ex ante, a coalition
must specify for every possible event the appropriate quantities
of commodities or net trades that each member receives or carries
out. In this case, an allocation used for blocking must guarantee
the coalition members a higher expected utility. Compared with
the concept of blocking an allocation ex post, that the ex ante
concept has caught on in the more recent literature because of its
compatibility with other models of general equilibrium.

2.8   Incentive Compatibility

Several models of the core of economies with asymmetric information have introduced the notion of *incentive compatibility* of
an allocation since the presence of private information may imply an incentive for an agent to misrepresent his information or

---

[10]'The plan of action *a* made initially for the whole future specifies for
each good and service the quantity that he will make available or that will
be made available to him, at each location, at each date, and at each event.
...Payment is irrevocably made although delivery does not take place, if
specified events do not obtain.' Debreu (1959), p.100.

his type. Therefore, only allocations which are incentive compatible are considered as possible elements of the core. Early statements of incentive compatibility constraints in the literature on the core include Berliant (1992), Boyd and Prescott (1986) and Marimon (1989). Allen (1991c) describes incentive compatibility of an allocation as follows: 'Incentive compatibility is taken to mean that players' strategies satisfy a self-selection constraint that the state-dependent net trade must give rise to allocations that are at least as good as those for the state-dependent net trade of another state of the world.' If 'at least as good' refers to conditional expected utility, this definition of incentive compatibility corresponds to the notion of Bayesian incentive compatibility as it is known from the theory of noncooperative games with incomplete information. She also defines a notion of *strong incentive compatibility* which applies to every state of the world separately. A similar concept of incentive compatibility has recently been employed by Vohra (1997).

A different notion of incentive compatibility has been introduced by Krasa and Yannelis (1994) and Koutsougeras and Yannelis (1994). They defined the concept of *coalition incentive compatibility*. This property of an allocation can explained intuitively as follows: In each coalition, the members reveal their information 'truthfully', i.e. if an allocation is coalitionally incentive compatible, then no coalition can improve upon this allocation by declaring a 'false' state of nature which the agents not belonging to this coalition cannot distinguish from the state which has occurred. A concept closely related, *coalition Bayesian incentive compatibility*, has been introduced by Hahn and Yannelis (1995) in a paper on implementation in economies with asymmetric information. The main difference being that in the former expected utility is used while the latter conditional expected utility is employed.

An important element in all formal definitions of incentive compatibility is the assumption that the preferences of the agents can be expressed by an expected utility function. It is unclear how this concept can be extended to general preference relations.

2.9   Definitions of the Core

The core of an economy comprises all feasible allocations that cannot be blocked by any coalition. As the literature, as described in sections 2.6 and 2.7, uses various definitions of feasible allocations and different concepts of blocking of an allocation are used, there is a corresponding variety of definitions of the core. There are three major alternative formulations of the core which differ, in particular, with respect to the assumptions about information exchange. Two of these were introduced by Wilson (1978) in the first model of the core of an economy with asymmetric information: the *coarse core* and the *fine core*. As Wilson assumes only the physical feasibility of allocations, these definitions differ only with regard to the communication system. As far as the coarse core is concerned, Wilson argues that the members of a coalition are not in a position to exchange information among each other, i.e. the null communication system is given. Reallocations of the initial endowments can only be carried out contingent on events which can be observed by all members of the coalition. The coarse core therefore comprises the set of all feasible allocations that cannot be blocked by any coalition, given the null communication system. In contrast, with the fine core, the assumption is that the coalition members are able to exchange their information without any restrictions, i.e. they can use the perfect communication system.[11] In this case, a coalition can attain all allocations that are compatible with the coarsest common refinement of the information of all coalition members. The fine core therefore includes all feasible allocations that are unblocked by any coalition, given the perfect communication system. As Wilson proceeds on the assumption of both an ex post utility maximization and an ex post blocking. As a consequence, the types of the core he considers are also ex post concepts.

The definition of the core used by Srivastava (1984a, 1984b) to

---

[11]Note that Wilson does not assume that a coalition has to use a certain communication system, but that this only represents the upper bound for the exchange of information.

a large extent coincides with Wilson's definition of the fine core. He also considers all allocations which are physically feasible as feasible for the economy. As already mentioned in section 2.5, Srivastava assumes that the information of each agent can only improve, but not deteriorate, through communication. Further, the maximal information can only be as fine as the 'pooled' information of all agents. This corresponds to Wilson's assumption that agents, given a perfect communication system, *can* but are *not obliged* to exchange their information. Srivastava's definition corresponds to Wilson's fine core except for the difference that Srivastava assumes an ex ante concept of blocking while Wilson uses an ex post concept: All physically feasible allocations that cannot be blocked by any coalition are elements of the fine core of an economy.

An alternative definition of the core was proposed by Yannelis (1991). Since he rules out the possibility of information exchange, each agent retains his individual information. Feasible allocations must be compatible with this information. The *private information core* introduced by Yannelis therefore comprises all allocations that are feasible given this assumption and that are at the same time unblocked by any coalition, where each coalition member retains his individual information. With reference to Wilson's work, Yannelis also considers the *coarse* and *fine core* with the corresponding assumptions concerning the exchange of information and the blocking of an allocation by a coalition. However, in this definition the possibilities of information exchange have *no influence whatsoever* on the set of feasible allocations. Only the allocations which are compatible with the individual information of the agents are feasible. This is a very restrictive assumption, particularly in the definition of the fine core.

The same requirements are maintained in the work of Koutsougeras and Yannelis (1994), in which these three types of the core are analyzed. Furthermore, they discuss two additional concepts of the core, *the strong coarse core* and *the weak fine core*, which differ with respect to the definition of feasible allocations. With the strong coarse core, the condition for the feasibility of an allocation is its compatibility with the events observed by *all* agents in the economy *jointly*. The strong coarse core thus comprises all allocations which are feasible in this sense and are unblocked by any coalition, when the coalition members can carry out their reallocations only contingent on events observed by all agents together. As far as the weak fine core is concerned, the condition for the feasibility of an allocation is modified in the following way: All allocations which are measurable with respect to the 'pooled' information are considered attainable. The 'pooled' information is the coarsest common refinement of the information of *all* agents in the economy. The weak fine core thus includes all feasible allocations which are unblocked by any coalition when the coalition members can exchange their information among each other without restrictions.

The definition of the core used by Allen (1991a, 1991b, 1992, 1994a) is closely related to the cases analyzed by Yannelis (1991) and Koutsougeras and Yannelis (1994). Because, however, in Allen's study the feasible allocations depend on the given information rule, coarse and fine core in her definition do not correspond with the definitions of of the coarse and fine core as described by Yannelis (1991) and Koutsougeras–Yannelis (1994). While Yannelis and Koutsougeras as well as Yannelis state that the feasible allocations must be measurable with respect to the individual information of the agents in both the coarse and the fine core, in Allen's models the set of feasible allocations changes with the information rule: if no information can be exchanged, feasible allocations are generated by net trades that are compatible with the information the agents can employ as members of the grand coalition. In other words, for an allocation to be feasible, the net trades have to be measurable with respect to events which are observed by all agents together. Allen's definition of the coarse core is therefore equivalent to the definition of the strong coarse core of Koutsougeras and Yannelis. Similarly, Allen's definition of the fine core corresponds to the definition of the weak fine core of Koutsougeras and Yannelis. As with the private information core rules out information exchange between the agents, the same sets of feasible allocations result in the models of Allen and Koutsougeras and Yannelis. Stated differently: With respect to the private information core, the definitions correspond to each other.

Compared to the other models, Allen's approach has the advantage that the description of feasible allocations takes the information rule into consideration. Furthermore, such an information allows for a compact description of the blocking of an allocation by a coalition. Thus, her approach offers a simple possibility – by a modification of the information rule – to analyze other types of the core of an economy than those described here.

There are also attempts to characterize the core of an economy with asymmetric information by a set of axioms, an approach which has been followed by Lee (1998). Starting from the axiomatization of the core of a game without transferable utility as introduced by Peleg (1985), Lee states five axioms which are known to be satisfied by the core of economies with perfect information. Lee is able to show that for economies with asymmetric information the coarse core is also characterized by these axioms.

In addition, Lee has defined an alternative core concept, the *coarse+ core*. In this concept, the blocking of an allocation by a coalition is defined as follows: A coalition will block an allocation if there is a subcoalition which could propose another allocation that would make all members of the coalition better off, i.e. if the subcoalition could make a dominant offer. The coarse+ core consists of all allocations which are unblocked in this sense. Lee provides four axioms and shows that the coarse+ core is characterized by these axioms.

Recently, some other definitions of the core of an economy with asymmetric information have been proposed. Starting from the communication process between the members of a coalition as described in section 2.5, Volij (1997) defines the *core with endogenous communication*. He shows that this core lies in between the coarse and the fine core as defined by Wilson (1978). He also has introduced the concept of the *internally consistent core*. Assuming that there is no exchange of information between the members of a coalition, a coalition can improve upon a given allocation only with an allocation contingent on an event observable by all members of the coalition. Such an improvement is *internally consistent* if there is no subcoalition who would block this improvement by another internally consistent allocation. The allocations for which this condition hold are elements of the internally consistent core.

Most models of the core of an economy allow only for one round of trading, i.e. agents agree upon state contingent net trades ex ante which are carried out after a state of the world has realized. It could be argued that the agents might have an incentive to engage in a second round of trading after the ex ante contracts have been carried out. This idea has been formalized by Koutsougeras (1998) who introduced the *two stage core*. Ex ante trades generate an allocation ex post which is taken to be the vector of endowments in a second round of trading. An allocation is in the two stage core if the allocation is in the usual core and if there is no coalition who could find ex ante net trades among its members that would generate a higher expected utility for each member than the given ex post allocation. The two stage core is a quite general concept as it can in principle be defined for all information rules, e.g. if the private information rule is given, the private two stage core results.

Taking into account problems of incentive compatibility as a defining characteristic of core allocations, concepts of the *incentive compatible core* have been introduced by Allen (1991, 1992, 1995) and by Vohra (1997). In both models the incentive compatibility constraints restrict the players' strategies in such a way that no allocation in the incentive compatible core gives a player an incentive to lie.

2.10   Existence of Core Allocations

As described in the previous sections, the studies of the core of an economy with asymmetric information vary both according to the required measurability conditions with respect to the initial endowments, allocations and net transactions, and the definition of feasible allocations. It is therefore not surprising that – depending on the combination of the various assumptions – diverse and sometimes contradictory results about the existence of core allocations indexallocation have been derived.

Thus, Wilson (1978) obtained the result that the coarse core is never empty, while the fine core can be empty. His proof of the existence of coarse core allocations is based on the construction of a cooperative game associated with an economy. Unlike the usual procedure of identifying the set of agents in the economy with the set of players in a cooperative game, Wilson models an agent by *several* players. Here, a player is given by an ordered pair, consisting of an agent and an element of the agent's informational partition. Wilson interprets this approach as follows: 'The substance of this argument is merely the observation that each agent can wear several hats in the negotiating process, or possibly he can delegate responsibility to subordinates, one for each event in his informational partition, to whom he confers responsibility in that event'[12] He thus shows that the game constructed in that way is balanced and the core of the game is therefore nonempty. Wilson provides an example to show that the fine core, in contrast, can be empty. As the cause for the nonexistence of allocations in the fine core, Wilson points out that a communication system merely represents the upper bound for the exchange of information. A coalition can use the perfect communication system to block an allocation, but is not obliged to do so. In his example, the only candidate for an allocation in the fine core is the initial endowment. However, the grand coalition blocks this by using the null communication system.

Yet Srivastava (1984a) was able to demonstrate that the nonexistence of allocations in the fine core – in contrast to Wilson's conjecture – is not due to the communication system, but to Wilson's assumptions of measurability. While Wilson requires only

---

[12]Wilson 1978, p.814

the measurability of the *allocations* with respect to the information, Srivastava also requires the compatibility of the and thus implicitly the net trades with the agents' information. If the initial endowments are measurable with respect to the individual information, then, according to Srivastava, the fine core is nonempty. His proof is based on the construction of an auxiliary economy with perfect information which is associated with the original economy. He shows that this economy possesses a nonempty core and that allocations in the core of the auxiliary economy are also elements of the core of the original economy with asymmetric information. However, Srivastava's claim that the measurability of the initial endowments with respect to information are necessary for the existence of fine core allocations is incorrect: as Allen later demonstrated, it is not the measurability of initial endowments that is decisive for the existence of allocations in the fine core, but the measurability of the net trades of each agent.

However, it is not only the measurability of the initial endowments, net trades and allocations that is important for the existence of fine core allocations. In addition, the definition of feasible allocations plays a crucial role. This is evident from the model of Koutsougeras and Yannelis (1994). Whereas in Srivastava's model, all physically feasible allocations are also feasible, Koutsougeras and Yannelis require the compatibility of feasible allocations with the individual information of the agents. In this case, the fine core is – contrary to Srivastava's result – generally empty. Koutsougeras and Yannelis provide an example which illustrates this point. Here, all allocations that satisfy this measurability condition can be blocked by coalitions using their 'pooled' information. However, these allocations are incompatible with the individual information of the agents and are therefore not feasible. It becomes obvious that the condition of feasibility of allocations is highly significant for the existence of fine core allocations when this type of core is compared with the weak fine core, which differs from the fine core only in the definition of feasibility of allocations: In this case, feasible allocations have to be compatible with the 'pooled' information of all agents. Koutsougeras and Yannelis now show that, in contrast to the fine core, the weak fine core is never empty.

A similar result applies to the coarse core and the strong coarse core, which also differ only in the definition of feasible allocations. For the coarse core, feasible allocations must be compatible with the initial information of the agents, while the strong coarse core requires the measurability of feasible allocations with respect to the finest common coarsening of all partitions. Koutsougeras and Yannelis are able to show that the coarse core is never empty, while the strong coarse core can be void. The latter is shown with the help of an example. Here, the information of the agents is chosen in a way that the finest common coarsening of all partitions is the trivial partition, i.e. the one that contains only one event. However, each individual agent has access to finer information. For the feasible allocations, this means that each agent has to consume amounts of commodities that are independent of the states of nature. But if the initial endowments are compatible with the individual information of the agents, but not with the finest common coarsening, then no allocation exists that guarantees every agent in every state the same consumption and gives a higher utility than the initial endowments. However, the initial endowments are not feasible because they are not measurable with respect to the finest common coarsening of the information. Therefore, the strong coarse core is empty.

This gives rise to the question whether a definition of feasible allocations that might imply infeasibility of the initial endowments is adequate. Considerations about plausibility give rise to doubts about the usefulness of such a feasibility condition. The autarky solution in which no transaction is carried out and every agent consumes his initial endowment, should certainly be considered as feasible for the economy.

The central result in the work of Yannelis (1991) and Koutsougeras and Yannelis (1994) concerns the existence of allocations in the private information core. The proof relies on the well-known theorem by Scarf (1967) about the existence of core allocations. However, since they proceed on the assumption of an infinite set of states and hence an infinite–dimensional commodity space, methods from the theory of Banach lattices have to be employed to check whether the assumptions of Scarf's theorem are satisfied. As the set of the feasible allocations is the same for the private information core and the coarse core, the former is nonempty and the latter is a superset of the private information core, the existence of coarse core allocations follows immediately.

As explained in section 2.9, the definitions of the weak fine core, the strong coarse core and the private information core of Koutsougeras and Yannelis correspond to the definitions of the fine core, coarse core and the private information core of Allen (1991a, 1991b, 1992, 1994a). Therefore, in Allen's models, the same statements concerning existence apply in principle for these types of the core. The weaker measurability requirements in Allen's models, i.e. only with respect to the net trades, has no consequence with regard to the existence theorems. As Allen requires only the measurability of the net transactions, she is able to guarantee that the initial endowments are always feasible. However, even with this weaker measurability condition, the coarse core according to Allen's concept is generally empty, because here often only net trades are allowed which are independent of the states of nature. Thus, the only possible candidate for a core allocation is the autarky solution. This allocation can, however, be blocked by coalitions which have access to better information than the grand coalition.

By using information rules to describe the exchange of information between the members of a coalition, Allen points out an interesting relationship between the properties of the information rules and the existence of core allocations: for all information rules with which the agents' information becomes finer as the size of the coalition increases, the core is nonempty.[13] The reason for this is that, for such information rules, the finest information that an agent can obtain is that which is available to him as a member of the grand coalition. All net transactions that can be carried

---

[13] Allen refers to such information rules as 'nested'. See Allen (1994a), p.12.

out by smaller coalitions are therefore measurable with respect to the information of the grand coalition, and therefore generate feasible allocations. However, with information rules which do not possess this property, smaller coalitions may have access to finer information than the grand coalition and hence may carry out net trades that are not compatible with the information of the agents as members of the grand coalition. These cases emphasize the special role of the grand coalition in the definition of feasibility.

With respect to some of the recently introduced concepts of the core, i.e. the core with endogenous communication, and the two stage core the following results about the existence of core allocations have been derived: Since the core with endogenous communication as introduced by Volij (1997) is closely related to the fine core as defined by Wilson (1978), the opportunities of communication may imply that the core is empty. With respect to the internally consistent core, Vohra does not show that this core is nonempty. The two stage core as defined by Koutsougeras (1998) however can be shown to be nonempty provided that the private information rule applies. However, no results concerning the existence of allocations in the coarse+ core or in the internally consistent core have been provided yet.

If incentive compatibility considerations are taken into account, Allen (1991c) shows that the incentive compatible core is generally empty. The reason for the emptiness of the core can be explained as follows: If incentive compatibility constraints are taken into account, the game generated by an economy with asymmetric information is in general not balanced since convex combinations of incentive compatible allocations are generally not incentive compatible. A similar result has been derived by Vohra (1997). However, Allen proves that the core is nonempty if randomization over state–dependent incentive compatible allocations is allowed for. The resulting allocation is ex ante incentive compatible and the randomization convexifies the set of incentive compatible allocations which implies that the game generated by the economy is balanced.

2.11   Properties of Core Allocations

While the earlier works on the core of an economy with asymmetric information are largely confined to statements about the existence of core allocations, properties of these allocations have been investigated in more recent contributions (Allen (1991a, 1991b, 1994a), Koutsougeras and Yannelis (1994)). They confirmed that there is a close relationship between the various possibilities of information exchange and the size of the core: given the individual information, the core of an economy is the smaller, the better are the possibilities of information exchange. The reason is that a coalition which can employ a finer information has more possibilities to block any given allocation, because the measurability conditions on the enforceable allocations are less restrictive. The coarser the information, the fewer the allocations that a coalition can use for blocking.

Intuitively, one would thus suspect that there is an inclusion relationship between the various types of core. Koutsougeras and Yannelis show in their model that the coarse core contains the private information core. The fine core, on the other hand – even if it is nonempty – is a subset of the private information core. This inclusion with respect the three types of the core holds true in their model because the set of feasible allocations is independent of the possibilities of information exchange between the agents.

Since in Allen's model the set of feasible allocations depends on the information rule, there is no similar inclusion between the coarse, the private information, and the fine core. In the coarse core all allocations are considered as feasible that can be achieved by net trades which are compatible with the finest common coarsening of the information of all agents in the economy. With respect to the private information core however, it is the individual information which determines feasibility, while for the fine core it is the coarsest common refinement of the information of all agents. Thus, the case might arise that there are allocations in the fine core that are not feasible given the 'private' or 'coarse' information rule. This property corresponds to a phenomenon frequently observed in the partial equilibrium literature about informational problems: When information is asymmetricly distributed, the 'first best' solution might not be not feasible. If, however, for two different information rules the information of the grand coalition is the same and if the information every agent in every coalition under one rule is finer than under the other, Allen proves an inclusion relationship between the respective cores. Since in this case the set of feasible allocations is the same, the core with the 'coarser' information rule contains the core with the 'finer' information rule.

With respect to incentive compatibility, Allen could show that the incentive compatible core is nonempty, provided that a randomization of incentive compatible allocations is allowed for. An incentive compatible allocation in the core is given by a probability distribution over state contingent transactions. This implies that the resulting core allocations are feasible only on average, i.e. it might happen that if a certain state realizes the allocation is not feasible.

Koutsougeras and Yannelis could show that allocations in both the coarse and the private information core are coalitionally incentive compatible, on condition that the preferences of the agents are monotonic. In these cases, information will be truthfully revealed in the core. They provide an example which shows that allocations in the fine core – if it is nonempty – are generally not coalitionally incentive compatible. Moreover, in the private information core, agents who possess finer information than others can profit from this informational advantage. This means that, because of his finer information, an agent can consume larger amounts of commodities than an otherwise identical agent with coarser information. Koutsougeras and Yannelis present an example which shows that, even when an agent has no initial endowment of commodities, he ends up with positive amounts of commodities in a private information core allocation. This is why they consider the private information core a suitable concept for the characterisation of allocations in the core of an economy with asymmetric information, since it takes this informational advan-

tage into account. The coarse core also contains only coalitionally incentive compatible allocations, however, they consider it as
'too large', because it covers all individually rational and Pareto–
efficient allocations.

2.12   Replica Theorems and Large Economies

In recent years, several contributions have been published in the
field of implementation theory which investigate the relationship
between the impact of asymmetric information on the efficiency of
allocations and the size of an economy (e.g. Gul and Postlewaite
(1992)).   It could be shown that, when an economy is replicated,
the impact of asymmetric information on the efficiency of allocations becomes less and less important and disappears in the limit.
In other words: In large economies, asymmetric information is
irrelevant.

While some papers on the core of an economy with asymmetric
information, as discussed in section 2.6, indicate that the set of
physically feasible allocations, or those which are feasible with
perfect information, becomes smaller. because agents are asymmetrically informed. Yet the question of the impact of informational problems in the context of the size of an economy has so
far been analyzed only by Srivastava (1984b). Here, a given economy is 'expanded' by replication. A replication of the economy
means that all agents, including their preferences, initial endowments and information, are duplicated. Every agent in the replicated economy is therefore of a certain type, and agents of the
same type are indistinguishable. Furthermore, the agents' information and the possibilities of information exchange are such that
the 'pooled' information of all agents in the original economy is
perfect.   On condition that the information of each agent does
not become coarser than his individual information through the
exchange of information, Srivastava shows that the core of the

economy shrinks when it is replicated, and converges to the core of an economy with perfect information. In other words: The effects of asymmetric information in large economies are negligible; asymmetric information does not cause any inefficiencies. However, for other types of the core, such as for example the coarse or private information core, the relationship between the size of an economy and the effects of asymmetric information has not yet been analyzed. It is therefore unclear whether Srivastava's statement also applies to the case of restricted information exchange. If incentive compatibility is taken into account, Allen (1994b) could show that for large replica economies the approximate core is nonempty and contains at least one equal treatment allocation, provided that randomization over state–contingent net trades is allowed for. The approximate core contains all those incentive compatible allocations that are feasible on average and that no coalition can improve upon by a discrete positve amount for each of it's members.

Allen has extended her analysis of economies with asymmetric information an economy to the case of a continuum of agents and incentive compatibility constraints (Allen (1995)). As mentioned above (see section 2.10), incentive compatibility constraints generally lead to nonconvex sets of attainable payoffs and incentive compatible core allocations generally don't exist. Using a theorem by Yamazaki (1978) about the existence of a competitive equilibrium with nonconvex consumption sets, Allen proves that in an exchange economy with a continuum of agents the incentive compatible core is nonempty provided that a mild assumption about the dispersion of the agents' enddowments is satisfied.

Chapter 3

Exchange Economies

The model of an exchange economy with asymmetric information as developed in this chapter differs from the approaches discussed so far in two important respects: The modeling of information and the definition of feasible allocations. *Information* is considered as a constituent part of an economy and not just as a restriction on the possible consumption plans as usually done in the literature. This allows for taking into account both aspects of information simultaneously, as mentioned in the introduction: Information as a commodity and information as a characteristic of an agent. Modeling information in this way is analogous to the treatment of uncertainty in models of general equilibrium, as e.g. Debreu (1953). He suggested to define commodities not only with respect to physical properties, time and place but also with respect to the state of nature in which the commodities are available. The definition employed here extends this approach by differentiating commodities with respect to information.

Feasible allocations are determined by the maximum amount of information available in the economy. This maximum amount of information depends on the individual information of the agents in the economy on the one hand, and on the possibilities of exchanging information on the other. While the literature considers the information of a certain coalition structure as determinant for feasibility – Allen (1991a, 1991b, 1994a) used the coarsest coalition structure, i.e. the grand coalition, Yannelis (1991) and Koutsougeras and Yannelis (1994) the finest coalition structure, i.e. the set of singleton coalitions – the definition employed here takes all coalition structures into account. Thus, it is independent of an arbitrary coalition structure. The maximum amount of information is therefore not necessarily the information the agents can employ as members of the grand coalition. Conceptually, the approach is similar to the one used in the theory of coalition production economies (see Böhm (1974a, 1988)). In these models the set of feasible allocations can be larger than the set of allocations the grand coalition is able to enforce. The definition of feasibility used here, takes into regard not only the endowment of commodities in the economy but also the 'endowment' of information.

The chapter is organized as follows: In the first section the model of an exchange economy with asymmetric information is developed. The information of an agent is described by a partition of a finite set of possible states of nature. Information is considered as part of the commodity space and thus as part of the agents' consumption sets. Section 3.2 introduces the concept of an information rule as developed by Allen (1991a). Some information rules are illustrated by examples. The cooperative games generated by those exchange economies are analyzed in section 3.3. It is shown that – under the usual assumptions – these games are balanced. Thus these games as well as the underlying economies have a nonempty core. Some extensions of the basic model, i.e. the problem of incentive compatibility, sequential trade and the question of information and learning are considered in section 3.5. The last section summarizes and discusses the main results.

3.1 Exchange Economies with Asymmetric Information

In the following section the central parts of an exchange economy with asymmetric information are discussed.

3.1.1 Agents and Their Information

We consider an economy with a finite set of agents denoted by I where a single agent is denoted by $i \in I$. The information of agent i is modelled in the usual way, i.e. as it has been done in the literature e.g. by (1968), Wilson (1978) or Allen (1991a, 1991b, 1992, 1994a).

Let $(\Omega, \mathcal{P}(\Omega))$ be a measurable space where the set ω is assumed to be finite.[1] The elements $\omega \in \Omega$ are called states of the world or states of nature. A state of the world $\omega \in \Omega$ is a complete description of all parameters relevant to a given economic model. These elements could include e.g. the weather or, on an individual level, the state of an agent's health. The set $\mathcal{P}(\Omega)$ denotes the power set of Ω, i.e. the sigma algebra of $\mathcal{P}(\Omega)$-measurable events.

The set of all partitions of Ω is denoted by P^*, and the partition which contains only the singleton subsets of Ω by Ω^*. The information of an agent $i \in I$ is characterized by a partition $P_i \in P^*$. A tuple $(P_i)_{i \in I}$ is called the information structure of the economy. The element of a partition P_i which contains the state ω is given by $P_i(\omega)$.

A function f with domain Ω is called P_i–measurable if it is measurable with respect to the sub–sigma algebra $\sigma(P_i)$ generated by the partition P_i. This sub–sigma algebra is the smallest sigma algebra of Ω containing all $P_i(\omega)$ with $\omega \in \Omega$. A function f is thus measurable with respect to P_i if it is constant on the elements of the partition P_i.

---

[1] The assumption of the finiteness of Ω is not necessary but facilitates the mathematics without reducing the economic content of the model.

Using this description of information, the following concepts can
be defined.

Definition 3.1 *An information P is called* finer *than an information P' if every element of P is a subset of an element of P',
i.e. $P(\omega) \subset P'(\omega)$ for all $\omega \in \Omega$. In this case the information P'
is called* coarser *than P.*

Definition 3.2 *An information P_i of agent i is called* perfect *or*
complete *if $P_i = \Omega^*$. An information structure of the economy is
called perfect if the information of each agent is perfect.*

Definition 3.3 *An information structure of the economy $(P_i)_{i \in I}$
is called* symmetric *if $P_i = P_{i'}$ for all $i, i' \in I$.*

Definition 3.4 *An information structure of the economy $(P_i)_{i \in I}$
is called* asymmetric *if there are at least two agents i, i' with
$P_i \neq P_{i'}$.*

While the definition of perfect information may refer only to the
information of one agent, the concepts of symmetric or asymmetric information are defined only for at least two agents. These
definitions imply the following statements: If the information in
the economy is perfect, it will also be symmetric, and if the information in the economy is asymmetric, then it cannot be perfect.
The converse of these statements is generally not true.

3.1.2   Commodities, Agents, and
Consumption Plans

The set of physical commodities in the economy is given by a finite
set L. A single commodity is denoted by $l \in L$. The physical
commodity space is given by $\mathbb{R}^L$, the L–dimensional Euclidean
space. The contingent commodity space is given by $Map(\Omega, \mathbb{R}^L)$,

the set of all mappings from the set of states of nature Ω to $\mathbb{R}^L$. Clearly, the set $Map(\Omega, \mathbb{R}^L)$ can be identified with the space $\mathbb{R}^{L\Omega}$. This identification also determines the topology of $Map(\Omega, \mathbb{R}^L)$.

Before defining the other constituents of the economy, it is expedient to introduce the following notation.

Definition 3.5 *For $P \in P^*$ the set $Map_P(\Omega, \mathbb{R}^L)$ denotes the following subspace of $Map(\Omega, \mathbb{R}^L)$:*

$$Map_P(\Omega, \mathbb{R}^L) := \left\{ f : \Omega \to \mathbb{R}^L \mid f|_{P(\omega)} \ constant \right\}.$$

The space $Map_P(\Omega, \mathbb{R}^L)$ consists of the P-measurable mappings $f : \Omega \to \mathbb{R}^L$, i.e. the mappings which are constant on elements of P.

The literature on the core of an economy with asymmetric information as discussed in chapter 2 has assumed that the information of an agent has an influence only insofar as it restricts the set of possible consumption plans. The approach used here differs from the literature as it considers information as a constituent part of the commodity space and thus of the economy. The approach is similar to the formulation of uncertainty where a commodity – besides place and time of availability – is determined also by the state of the world in which it is available.[2] In this model physical commodities cannot be considered independently of the states of nature. Taking into regard information, it is natural to assume that state contingent commodities cannot be considered independently of information. Thus, a *generalized commodity space* $Map(\Omega, \mathbb{R}^L) \times P^*$ is considered. In this commodity space, commodities are ordered pairs, consisting of a state contingent commodity and an information. Intuitively, this definition implies that information becomes part of the description of a state contingent commodity.

To illustrate this idea by a simple example, consider two different states of the world. In one of these, agent i is infected with a

---

[2]see Debreu (1953), p.120.

virus. In the other, he is not infected. Consider two possible information for agent i: P_i which allows him to distinguish these states, and P_i' which does not allow for a distinction. Suppose the disease can be cured by a remedy. Clearly, the remedy with information P_i' is not the same commodity as the remedy with information P_i.

This approach could in principle be extended to the case of an infinite number of states of the world. However, this would require first, to define a topology on the information sets (in this case the sub–sigma algebras of Ω) and secondly, to deal with an infinite dimensional commodity space. Both problems have been considered in the literature. Topologies on information sets have been introduced e.g. by Allen (1983), Cotter (1986), and Stinchcombe (1990) by defining a metric on the set of sub–sigma algebras.

Cooperative models with infinite dimensional commodity spaces have also been discussed e.g. by Yannelis (1991) as well as Koutsougeras and Yannelis (1994). However, to analyze such an economy, methods from measure theory or from the theory of Banach lattices have to be employed.

In the model examined here however, only finitely many states of the world are considered. This keeps matters as simple as possible while retaining the economic contents of the model. Therefore, the commodity space is finitely dimensional and information can easily be introduced as part of the definition of a commodity. Of course, this amounts to endowing the information structures, i.e. the partitions of Ω with the discrete topology.

As to the other constituents of the economy, each agent $i \in I$ is characterized by a nonnegative initial endowment $(e_i, P_i^0) \in Map(\Omega, \mathbb{R}_+^L) \times P^*$, an element of the generalized commodity space. Here e_i denotes his endowment with state contingent commodities and P_i^0 denotes agent i's endowment of information, or his initial information. It is assumed that the physical endowment e_i is compatible with P_i^0, since otherwise the agent could infer additional information from his initial endowment of contingent commodities.[3]

The *consumption set* of agent i is defined by

$$X_i := \left\{ (x, P) \in Map(\Omega, \mathbb{R}^L_+) \times P^* | x - e_i \in Map_P(\Omega, \mathbb{R}^L) \right\}.$$

Obviously, the relation $(e_i, P^0_i) \in X_i$ holds, since $e_i - e_i$ is compatible with every information.

A *consumption plan* for an agent i is an ordered pair (x, P) where the first component is a contingent commodity, and the second denotes his information. The agent's consumption set consists of all consumption plans that can be realized by net trades compatible with the information P. Notice that this holds for *all* conceivable partitions P. That is, a plan (x, P) for an arbitrary P is in the consumption set if $x - e_i$ is constant on the elements of P. An agent cannot consume contingent commodities independently of his information: If he is unable to discriminate between two states of nature, then consumption plans achieved by net trades that differ in these two states are not in his consumption set.

Notice that our definition of the consumption set allows for an agent's information to change. It departs from the one usually employed in the literature in that information is modelled as a constituent part of the consumption plans.

The main difference between the definition of consumption sets and consumption plans as used in the literature and the definition employed here can be characterized as follows: While the former considers information only as a restriction on the consumption set, the latter makes information a fundamental part of the consumption set and thus of the economy.

An *allocation* is an I–tuple of consumption plans $((x_i, P_i))_{i \in I}$ with $(x_i, P_i) \in X_i$. Thus, an allocation consists not only of contingent commodity vectors, i.e elements of $Map(\Omega, \mathbb{R}^L_+)$, but also comprises the agents' information.

---

[3]For the results, the assumption of e_i being compatible with P^0_i is not necessary. It turns out that the measurability of the net trade $x - e_i$ is sufficient.

3.1.3　Preferences

It is assumed that an agent i's preferences can be represented by a continuous utility function

$$U_i : X_i \to \mathbb{R}.$$

This utility function is assumed to possess the following invariance property:

$$U_i(x, P) = U_i(x, P') \quad \text{for} \quad (x, P), (x, P') \in X_i.$$

In words: Two consumption plans which differ only with respect to the information yield the same utility. The information does not generate any utility. Note that this invariance property applies only to consumption plans generated by net trades compatible with both P and P'. Obviously, this equation implies:

$$U_i(x, P) = U_i(x, \Omega^*) \quad \forall (x, P) \in X_i.$$

This means that consumption plans achieved by net trades compatible with the information P can also be generated by net trades compatible with perfect information since each consumption plan (x, Ω^*) is always an element of the consumption set.

Setting

$$\tilde{U}_i(x) := U_i(x, \Omega^*)$$

yields

$$U_i(x, P) = U_i(x, \Omega^*) = \tilde{U}_i(x) \quad \forall (x, P) \in X_i.$$

This connects a utility function defined on subsets of the generalized commodity space to a utility function defined on contingent commodities only: The utility function U_i is mainly the trivial extension of the function $\tilde{U}_i$ which is defined only for contingent commodities. In this sense the utility function U_i is independent of the information. Exactly the form $\tilde{U}_i$ of the utility function results if information is perfect.

Further, the definition of $\tilde{U}_i$ implies that

$$U_i(x, P) \geq U_i(y, P') \iff \tilde{U}_i(x) \geq \tilde{U}_i(y) \quad \forall (x, P), (y, P') \in X_i.$$

That is, the utility generated by a consumption plan (x, P) is larger than or equal to the utility of a consumption plan (y, P') if and only if the utility of a contingent commodity vector x is larger or equal to the utility of a contingent commodity vector y, provided that the net trades $x - e_i$ and $y - e_i$ are measurable with respect to P and P', respectively.

Using $\tilde{U}_i$, the concept of (quasi)–concavity of a utility function $U_i(x, P)$ can be defined as follows:

Definition 3.6 *A utility function $U_i(x, P)$ is (quasi)-concave, if the utility function $\tilde{U}_i(x)$ is (quasi)-concave.*

For the case of linear preferences, the utility functions $\tilde{U}_i$ can be represented by an expected utility function of the form

$$\tilde{U}_i(x) := \sum_{\omega \in \Omega} u_i(x(\omega))\mu_i(\omega) \quad \forall x \in Map(\Omega, \mathbb{R}_+^L),$$

where $\mu_i(\omega)$ denotes the ('subjective') probability of state ω and $u_i : \mathbb{R}_+^L \to \mathbb{R}$. In addition, $\mu_i(\omega) \geq 0$ for all $\omega \in \Omega$ and $\sum_{\omega \in \Omega} \mu_i(\omega) = 1$ for all $i \in I$. The utility function U_i is of the form

$$U_i(x, P) = \tilde{U}_i(x) = \sum_{\omega \in \Omega} u_i(x(\omega))\mu_i(\omega).$$

In this case, the utility function U_i can be represented by a function u_i defined on the physical commodity space and a probability distribution μ_i over all states of the world.

Having defined all the constituents of an exchange economy with asymmetric information $\mathcal{E}$, this economy can now be written as

$$\mathcal{E} := \left(I, \left(X_i, U_i, (e_i, P_i^0) \right)_{i=1}^I \right).$$

For $|\Omega| = 1$, this definition corresponds to the definition of an exchange economy usually given in the literature (cf. e.g. Hildenbrand and Kirman (1988))[4].

3.2   An Exchange Economy with Information Rule

Up until now, the information of an agent has been given parametrically. Subsequently, we will allow for an agent's information to change if he joins a coalition.

3.2.1   Coalitions

A coalition is a set $S \subset I$. The agents in a coalition are ordered according to their order in I, thus each nonempty coalition inherits the order $i_1, \ldots, i_S$ induced by I. That is, $S = \{i_1, \ldots, i_S\}$ with $1 \le i_1 < i_2 < \cdots < i_S \le |I|$. The set of coalitions in the economy is given by $\mathcal{P}(I)$.

One could imagine that the members of a coalition are able to exchange or aggregate their individual information in a way that, by joining a coalition, each agent can employ a possibly different information. Reallocations of endowments can now be carried out

---

[4]Here $|\cdot|$ denotes the cardinality of a set.

on the basis of this modified information. The change of information is described by an exogenously given *information rule,* a concept proposed by Allen (1991) under the name of an *information sharing rule.*

3.2.2 Information Rules

An information rule assigns to each coalition and to each tuple of its members' information a new tuple of information. The i-th component of this new tuple denotes the information that coalition member i can use if he joins the coalition. Formally, an information rule is a set of $2^{|I|}$ mappings k_S, where k_S assigns to each $|S|$-tuple $(0 \leq |S| \leq |I|)$ of information another $|S|$-tuple. Obviously, $k_\emptyset = \emptyset$, i.e. $k_\emptyset$ is the empty mapping.

Definition 3.7 *An* information rule *for a coalition $S \subset I$ is a mapping*

$$k_S : (P^*)^S \to (P^*)^S$$

with the property $k_S = Id$ if $|S| = 1$, i.e. singleton–coalitions and agents can be identified. Here Id denotes the identity mapping. An information rule for the economy is thus a $2^{|I|}$–tuple $k = (k_S)_{S \in \mathcal{P}(I)}$ of information rules, one for each coalition. For $S \neq \emptyset$, k_S can be written as

$$k_S \left((P_i)_{i \in S} \right) = (P_i')_{i \in S}.$$

In words: As a member of coalition S, agent i with information P_i has access to the information P_i', $i \in S$. The information accessible to the i-th coalition member is denoted by $k_S^i \left((P_j)_{j \in S} \right)$, i.e. the projection on the ith component.

Intuitively, an information rule describes the information which the members of a coalition S can use for reallocating their endowments. It allows for a simple formal description of modified information due to coalition membership.

Denoting the set of all information rules by $\mathcal{K}$, we define a binary relation $\precsim$ (read: *finer as*) on $\mathcal{K}$ as follows.

Definition 3.8 *For any two information rules k and k', the relation $k \precsim k'$ holds if $\forall\, S \subset I$, $\forall\, i \in S$ and $\forall\, (P_{i_1}, \ldots, P_{i_S}) \in (P^*)^S$:*

$$k_S^i \left((P_j)_{j\in S}\right) \quad \text{is finer than} \quad k_S'^i \left((P_j)_{j\in S}\right)$$

according to Definition 3.1.

Thus, an information rule k is finer than an information rule k' if each element of an information induced by k is a subset of some element of an information induced by k', i.e. if the information for each agent is at least as fine under k as under k'. The relation $\precsim$ is reflexive and transitive, but not complete. Therefore, the set $\mathcal{K}$ is partially preordered by the relation $\precsim$.

Some other important properties of information rules have been introduced by Allen (1991).

Definition 3.9 *An information rule k is symmetric if $\forall\, S \subset I$, $\forall\, i, j \in S$*

$$k_S^i \left((P_j)_{j\in S}\right) = k_S^j \left((P_i)_{i\in S}\right).$$

In words: all members of a coalition can employ the same information.

Definition 3.10 *An information rule k is nested if for all $i \in I$ and all coalitions S and T with $i \in S \subset T \subset I$ holds*

$$k_T^i \left((P_j)_{j\in T}\right) \precsim k_S^i \left((P_j)_{j\in S}\right).$$

If a coalition increases in size, the information of each member will not decrease.

Definition 3.11 *An information rule k is bounded if for all $i \in I$ and all coalitions S*

$$k_I^i \left((P_j)_{j \in I} \right) \precsim k_S^i \left((P_j)_{j \in S} \right).$$

Boundedness of an information rule is thus simply nestedness applied to the grand coalition. Obviously, nested information rules are bounded but not vice versa.

To illustrate the concept of an information rule, we present three examples which correspond to three core concepts which have been analyzed in the literature. These are the *coarse*, *fine*, and *private information rule*.

Definition 3.12 *The* coarse information rule *The* coarse information rule $k^c := (k_S^c)_{S \in \mathcal{P}(I)}$ *is given by*

$$k_S^c \left((P_i)_{i \in S} \right) = (P_S^c)_{i \in S}, \text{ where } P_S^c := \bigwedge_{i \in S} P_i \quad \forall S \neq \emptyset.$$

Here $\bigwedge_{i \in S} P_i$ denotes the finest coarsening of the partitions, such that for all P_i, each element of P_i is a subset of an element of $\bigwedge_{j \in S} P_i$.

The coarse information rule describes the situation where the members of a coalition are unable to exchange information. Only reallocations contingent on events discernible by all members of the coalition are possible. It is important to note that this does not require the agents' 'forgetting' their initial information when joining a coalition. However, if information transmission is impossible, they might be unable to *use* their initial information in a coalition. The coarse information rule is symmetric, but it is neither nested nor bounded.

Definition 3.13 *The* fine information rule $k^f := (k^f_S)_{S \in \mathcal{P}(I)}$ *is given by*

$$k^f_S ((P_i)_{i \in S}) = (P^f_S)_{i \in S}, \text{ where } P^f_S := \bigvee_{i \in S} P_i \quad \forall S \neq \emptyset.$$

Here $\bigvee_{i \in S} P_i$ is defined as $\left(\bigvee_{i \in S} P_i \right)(\omega) := \bigcap_{i \in S} P_i(\omega)$ for all $\omega \in \Omega$.

The resulting information of the coalition members is the coarsest refinement of the P_i with the property that each element of $\bigwedge_{j \in S} P_i$ is a subset of elements of the P_i. The fine information rule describes the situation where the members of a coalition can exchange information effectively. In this case, the information that can be used for the reallocation of endowments will generally be finer than the individual's information. Notice that the fine information rule is symmetric as well as nested and bounded.

While the coarse and the fine information rules are both symmetric, the private information rule introduced by Yannelis (1991) allows for coalition members to be asymmetrically informed.

Definition 3.14 *The* private information rule $k^p := (k^p_S)_{S \in \mathcal{P}(I)}$ *is given by*

$$k^p_S ((P_i)_{i \in S}) = (P_i)_{i \in S} \quad \forall S \neq \emptyset, \quad i.e. \quad k^p_S = Id \ \forall S.$$

In words: The mapping k^p_S does not change the information of any coalition member.

Obviously, the following relation holds for the information rules mentioned above: $k^f \precsim k^p \precsim k^c$.

These three information rules are illustrated by the following simple example:

Example 3.1 Let the set of states of nature given by

$$\Omega = \{\omega_1, \omega_2, \omega_3, \omega_4\}.$$

Two agents, who can form a coalition $S = \{1, 2\}$ have the following initial information:

$$P_1^0 = \{\{\omega_1, \omega_3\}, \{\omega_2, \omega_4\}\},$$

$$P_2^0 = \{\{\omega_1, \omega_2\}, \{\omega_3, \omega_4\}\}.$$

If the fine information rule is given, the information of agents 1 and 2 as members of the coalition S are given by:

$$\begin{aligned} k_S^f(P_1^0, P_2^0) &= (\{\{\omega_1\}, \{\omega_2\}, \{\omega_3\}, \{\omega_4\}\}, \\ &\quad \{\{\omega_1\}, \{\omega_2\}, \{\omega_3\}, \{\omega_4\}\}). \end{aligned}$$

Under the coarse information rule they are

$$k_S^c(P_1^0, P_2^0) = (\{\{\omega_1, \omega_2, \omega_3, \omega_4\}\}, \{\{\omega_1, \omega_2, \omega_3, \omega_4\}\}),$$

while the private information rule leads to

$$k_S^p(P_1^0, P_2^0) = (\{\{\omega_1, \omega_3\}, \{\omega_2, \omega_4\}\}, \{\{\omega_1, \omega_2\}, \{\omega_3, \omega_4\}\}).$$

These rules correspond to the coarse, fine and private information core as considered in the literature. However, the concept of an information rule is much more general. In fact, *any* kind of change in the agents' information can be described by such a rule. An extreme case of an information rule is the *null information rule* k^n. Independently of the agents' initial information, this rule associates to each coalition member the information $P_i = \{\Omega\}$. Notice that this information rule differs from Wilson's null communications system, which corresponds to the private information rule.

For example, consider a trial, where all persons involved, judge, prosecutor, defense counsel, witnesses, and jury know exactly that the accused has committed the crime; there is conclusive evidence.

Nevertheless, it is possible that the accused will exit the law court as a free man. This will happen if the evidence has been gained in an unlawful manner and is not allowed to be used in the trial.

Using the concept of an information rule, we can now define an *exchange economy with information rule* as follows: An *exchange economy with information rule* $\mathcal{E}^k$ is a tuple

$$\mathcal{E}^k := \left(I, \left(X_i, U_i, (e_i, P_i^0) \right)_{i=1}^{I}, k \right),$$

where $\left(I, (X_i, U_i, e_i, P_i)_{i=1}^{I} \right)$ is an exchange economy and k is an information rule for this economy.

3.2.3   Feasible Allocations

As discussed in chapter 2, section 2.10, the definition of feasible allocations is of fundamental importance with respect to the existence of core allocations. Of course, feasible allocations have at least to be physically feasible. However, if the economy is also characterized by asymmetric information, a sensible definition of feasibility should also take into account the initial information of the agents as well as the possibilities of information sharing, i.e. the information rule.

In the survey on the literature (p.25ff) it has been pointed out that mainly two definitions of feasibility are employed: Allen (1991a, 1991b, 1994a) considers only those allocations as feasible which can be realized by net trades compatible with the information the agents can employ if they are members of the *grand coalition*. In the work by Yannelis (1991) and Koutsougeras and Yannelis (1994) the compatibility of the net trades with the *initial information* of the agents is assumed. Thus, both definitions use a certain coalition structure in their definition of feasibility: According to Allen's definition, *coarsest* coalition structure is decisive for feasibility, while Koutsougeras and Yannelis employ the *finest* coalition structure, i.e. the set of singleton–coalitions. In our opinion, it is not quite clear why feasibility should depend on a certain coalition structure. In the first place, the choice of one particular coalition as a determinant for feasibility seems rather arbitrary. There is no reason why any one coalition structure be *a priori* distinguished. Further, both definitions might lead to counterintuitive implications. According to the definition given by Koutsougras and Yannelis (1995), the possibility of information exchange cannot affect feasibility. In this model, the set of feasible allocations is independent of the information rule. This contradicts the intuition that the set of feasible allocations should increase if information sharing is possible. On the other hand, Allen's definition imposes implausible restrictions on the set of feasible allocations. For instance, in the case of the coarse information rule, it might happen that the members of the grand coalition can use no information at all. As a consequence, only the initial endowments might remain feasible even if smaller coalitions with finer information on the part of their members would be able to carry out reallocations of their endowments.

This leads to the conclusion that in economies with asymmetric information, an allocation should at least be physically feasible, but also depend on (i) the agents' initial information, (ii) the way agents can exchange information, i.e. the information rule, and (iii) should be independent of any particular coalition structure. One could imagine that feasibility be determined by the finest information an agent could possibly achieve in any coalition. However, since the relation 'finer' induces only a partial preorder on the set of all partitions, the finest information is generally not unique. There might be several finest information which are incomparable. Therefore, we propose the following aggregation: Given an information rule, determine for each agent all the information he can use in all coalitions, and then pool these information. Formally, the resulting information is defined as follows:

$$P_i^m := \bigvee_{S \ni i} k_S^i((P_j)_{j \in S}).$$

Thus, P_i^m describes the maximal information agent i could acquire (given an information rule) if he were able to join all coalitions simultaneously. This concept takes into account all possible coalition structures, as well as the agents' initial information and the information rule. The tuple $(P_1^m, \ldots, P_I^m)$ is the maximum amount of information in the economy, given the initial information of all agents and the information rule. Notice that this concept is used only to describe the set of allocations which can ideally be conceived of as feasible for the economy. It is not assumed that any agent has indeed access to this maximum amount of information.

Using the concept of the maximum amount of information, feasibility for a given information rule k is now defined as follows:

Definition 3.15 *An allocation* $((x_i, P_i))_{i \in I}$ *is called* k–feasible, *if:*

$$1. \qquad \sum_{i \in I} x_i = \sum_{i \in I} e_i$$

and

$$2. \qquad x_i - e_i \in Map_{P_i^m}(\Omega, \mathbb{R}^L) \quad \forall i \in I.$$

The set of k–feasible allocations is denoted by E^k.

Condition 1 of the definition refers to physical feasibility, while condition 2 describes informational feasibility. Thus, an allocation is feasible with respect to an information if it is achievable by net trades compatible with the maximum amount of information, given information rule k. If a given economy with information rule is considered, an allocation is called 'feasible' if it is k–feasible. In this case the set of feasible allocations is denoted by E.

We now provide an example to illustrate the differences between our definition of feasibility and the ones used by Allen (1991) and Koutsougeras and Yannelis (1993). This example also illustrates that the maximum amount of information does not necessarily coincide with what is usually referred to as 'perfect' or 'full' information in the literature.

Example 3.2 Consider an economy with $\Omega = \{\omega_1, \omega_2, \omega_3, \omega_4\}$, one physical commodity and three agents with initial information

$$P_1^0 = \{\{\omega_1, \omega_2\}, \{\omega_3\}, \{\omega_4\}\}$$

$$P_2^0 = \{\{\omega_1, \omega_2, \omega_3\}, \{\omega_4\}\}$$

$$P_3^0 = \{\{\omega_1\}, \{\omega_2\}, \{\omega_3\}, \{\omega_4\}\}.$$

The endowments are given by

$$(e_1(\omega_1), e_1(\omega_2), e_1(\omega_3), e_1(\omega_4), P_1^0) = (1, 1, 2, 0, P_1^0)$$

$$(e_2(\omega_1), e_2(\omega_2), e_2(\omega_3), e_2(\omega_4), P_2^0) = (1, 1, 1, 3, P_2^0)$$

$$(e_3(\omega_1), e_3(\omega_2), e_3(\omega_3), e_3(\omega_4), P_3^0) = (1, 1, 2, 0, P_3^0).$$

If the coarse information rule applies, the set of feasible allocations is determined by those allocations which are physically feasible and can be gained by net trades compatible with the initial information P_i^0, $i = 1, 2, 3$. For example, the allocation

$$(x_1(\omega_1), x_1(\omega_2), x_1(\omega_3), x_1(\omega_4), P_1^0) = (0.5, 0.5, 2, 1, P_1^0)$$

$$(x_2(\omega_1), x_2(\omega_2), x_2(\omega_3), x_2(\omega_4), P_2^0) = (1.5, 1.5, 1.5, 1, P_2^0)$$

$$(x_3(\omega_1), x_3(\omega_2), x_3(\omega_3), x_3(\omega_4), P_3^0) = (1, 1, 1.5, 1, P_3^0)$$

is feasible since the net trades as well as the resulting allocation is compatible with respect to the agents' initial information. According to Allen's definition, however, this allocation would not be feasible since the net trades

$$x_1 - e_1 = (-0.5, -0.5, 0, 1),$$

$$x_2 - e_2 = (0.5, 0.5, 0.5, -2),$$

$$x_3 - e_3 = (0, 0, -0.5, 1)$$

are not measurable with respect to the information accessible to the members of the grand coalition, i.e. $P_i^c = \{\{\omega_1, \omega_2, \omega_3\}, \{\omega_4\}\}$. In this case, the maximum amount of information is less than 'perfect' or 'full' information.

If the fine information rule applies, the maximum amount of information in the economy is the perfect or full information. Therefore, feasibility coincides with physical feasibility. In contrast, the definition used by Koutsougeras and Yannelis (1993) implies that all allocations with $x_1(\omega_1) \neq x_1(\omega_2)$ or $x_2(\omega_2) \neq x_2(\omega_3)$ are not feasible since they cannot be achieved by net trades measurable with respect to the information P_1^0 or P_2^0. This example shows that the model considered here is neither a special case of Koutsougeras and Yannelis (1993) nor of Allen (1991).

Notice that if the information rule is such that the information gets finer with increasing coalition size, our definition of feasibility is equivalent to the one proposed by Allen. In this case, the larger the coalition, the finer the information the members can employ for their transactions. Of course, the members of the grand coalition have access to the finest information. On the other hand, if information gets coarser with increasing coalition size, our definition of feasibility is equivalent to the one used by Yannelis: Joining a coalition does not generate any additional information; in general, the singleton coalitions determine the finest possible information.

Finally, in the case of perfect information, our definition reduces to physical feasibility: If $P_i^m = \Omega^* \, \forall i \in I$ holds, condition 2 of definition 4.6 imposes no informational restriction on the set of physically feasible allocations.

The following lemma states an important property of the set of feasible allocations.

Lemma 3.1 *The set of feasible allocations E is compact.*

Proof. By definition, the consumption set X_i is a subspace of $Map(\Omega, \mathbb{R}_+^L) \times P^*$. Obviously, the set $A := \{x \in Map(\Omega, \mathbb{R}_+^L) | x \leq \sum_{i \in I} e_i\}$ is closed and bounded and thus compact. The feasible allocations E are elements of $(A \times P^*)^I$. Since P^* is a finite set and A is compact, the set $A \times P^*$ and thus $(A \times P^*)^I$ is compact. Therefore, it suffices to show that E is closed in $(Map(\Omega, \mathbb{R}_+^L) \times P^*)^I$. Consider a sequence $g^n = (g_i^n)_{i \in I}$ in E with $g^n \to g = (g_i)_{i \in I} \in (Map(\Omega, \mathbb{R}_+^L) \times P^*)^I$, where $g_i^n = (x_i^n, P_i^n)$ and $g_i = (x_i, P_i)$. Since P^* is a finite and thus discrete set, it follows that $P_i^n = P_i$.

Since $\sum_I x_i^n = \sum_I e_i \, \forall n$ and $x_i^n \to x_i$, the equation $\sum_I x_i = \sum_I e_i$ holds. Further, $x_i^n - e_i \in Map_{P_i^m}(\Omega, \mathbb{R}^L)$, therefore also $x_i - e_i \in Map_{P_i^m}(\Omega, \mathbb{R}^L)$. Thus it is shown that g is an element of E. Therefore, E is closed. Being a closed subset of the compact space $(A \times P^*)^I$, E is also compact. $\qquad\square$

Having defined the set of feasible allocations, we now turn to the core of an exchange economy with asymmetric information.

3.2.4 The Core of an Economy with Information Rule

To define the core of an exchange economy, it is necessary to introduce the concept of a coalition's *blocking of an allocation*.

Definition 3.16 *A nonempty coalition $S \subset I$ k–blocks an allocation $f = ((x_i, P_i))_{i \in I}$ if there is an allocation $g = ((y_i, P_i'))_{i \in I}$ such that*

1. $\sum_{i \in S} y_i = \sum_{i \in S} e_i$,

2. $(P_i')_{i \in S} = k_S ((P_i^0)_{i \in S})$,

3. $U_i(y_i, P_i') > U_i(x_i, P_i)$, $\forall i \in S$.

An allocation $g = ((y_i, P_i'))_{i \in I}$ with properties 1 and 2 is called enforceable by S.

Analogously to the definition of feasible allocations, we just say 'blocking' if a given economy with information rule k is analysed. Notice that, since $y_i - e_i \in Map_{P_i'}(\Omega, \mathbb{R}^L)$ $\forall i \in S$, it follows that $y_i - e_i \in Map_{P_i^m}(\Omega, \mathbb{R}^l)$, because the information of agent i as member of coalition S, P_i', cannot be finer than his maximum amount of information P_i^m.

Condition 2 states that allocations which can be used by a coalition S to block a given allocation have to be achievable by net trades compatible with the information the coalition members can use. Put differently, the information the members of the coalition can employ to block an allocation depends on the agent's initial information as well as on the information rule. The better the possibilities of information sharing, the more allocations can be blocked and vice versa. Thus, the set for enforceable allocations generally varies with the information rule.

Using the definition of feasible allocations and the concept of blocking of an allocation , the core of an exchange economy with information rule k is defined as follows:

Definition 3.17 *An allocation $f = ((x_i, P_i))_{i \in I}$ is in the k–core of the economy if the following conditions are satisfied:*

1. *the allocation f is k–feasible;*

2. *there is no coalition which k–blocks the allocation f.*

We say 'core of an economy' if it is clear which information rule applies.

The impact of asymmetric information is illustrated by a simple example.

Example 3.3 Consider an economy with two agents, $i = 1, 2$ and two states of nature, i.e. $\Omega = \{\omega_1, \omega_2\}$. The agents' initial information is given by

$$P_1^0 = \{\{\omega_1\}, \{\omega_2\}\},$$

$$P_2^0 = \{\{\omega_1, \omega_2\}\}.$$

There are two state contingent commodities in the economy. The initial endowments of the agents are given by

$$(e_{11}, e_{12}, P_1^0) = (3, 4, P_1^0),$$

$$(e_{21}, e_{22}, P_2^0) = (1, 1, P_2^0),$$

where the first subscript refers to the agent and the second to the state contingent commodity. The agents have identical preferences which can be represented by the utility function

$$\tilde{U}_i(x_1, x_2) = x_1 x_2, \quad i = 1, 2.$$

If the agents can pool their information, i.e. if the fine information rule k^f applies, the maximum amount of information for each agent is given by

$$P_i^m = \{\{\omega_1\}, \{\omega_2\}\}, \quad i = 1, 2,$$

i.e. perfect or full information. Thus the set of feasible allocations comprises all allocations that satisfy the equations

$$x_{11} + x_{21} = 4,$$

$$x_{12} + x_{22} = 5.$$

In this case the feasible allocations are not restricted by any measurability constraints caused by imperfect information. The core of this economy contains all feasible allocations which are individually rational and Pareto–optimal, i.e. for which

$$x_{12} = 1.25x_{11}$$

and

$$\tilde{U}_1(x_{11}, x_{12}) \geq 12,$$

$$\tilde{U}_2(x_{21}, x_{22}) \geq 1.$$

Notice that the initial endowments are not in the core since the allocations would be blocked by the grand coalition. E.g. the Pareto–optimal allocation

$$x_{11} = 3,104, \quad x_{12} = 3.88,$$

$$x_{21} = 0,896, \quad x_{22} = 1.12$$

could be used for blocking. In this allocation, agent 1 ends up with a utility of 12.044 while agent 2 gets a utility of 1.004.

However, if either the private or the coarse information rule applies, the maximum amount of information for each agent is identical to his initial information. The set of feasible allocations is restricted to those allocations which can be reached by net trades compatible with this information, i.e. net trades which are constant. The reason is that agent 2 cannot discriminate between the two states of the world and is therefore unable to make state contingent transactions. The core allocation given above is thus not feasible. The private as well as the coarse core contain only the initial endowments. All other allocations would be blocked by agent 1 or agent 2.

This example illustrates that asymmetric information and incomplete information sharing might be the reason why there are no transactions between the agents. This phenomenon is analogous to the 'lemons problem' discussed by Akerlof. In his model asymmetric information may induce a breakdown of a market, or an inefficiently low level of transactions.[5]

3.3 The Game Generated by an Exchange Economy with Information Rule

In this section it will be shown that an exchange economy with information rule generates a cooperative game. For this purpose, a utility allocation is defined as follows:

Definition 3.18 *Let* $a = ((x_i, P_i))_{i \in I}$ *be an allocation. Then* $(U_i(x_i, P_i))_{i \in I}$ *is called the* utility allocation *corresponding to a.*

In general, a cooperative game (with non–transferable utility) is completely characterized by the set of players and the characteristic function. In our framework, however, the set of feasible utility allocations has to be taken into account, since feasibility is not determined by what the grand coalition can achieve. For instance, in the case of the coarse information rule, the set of feasible utility

---

[5]see Akerlof (1970).

allocations may be larger than the set of allocations enforceable by the grand coalition because, in general, the members of the grand coalition possess a coarser information than the members of smaller coalitions. A similar phenomenon occurs in the theory of coalition production economies, where the set of feasible allocations may be larger than the set of allocations enforceable by the grand coalition.

Thus, in the framework considered here, a cooperative game with non–transferable utility in characteristic function form is given by a tuple (I, F, V), where I denotes the set of players, F is the set of feasible utility allocations, and V is a correspondence $V : \mathcal{P}(I) \to \mathbb{R}^I$. It is assumed that V has the following properties: $V(\emptyset) = \{0\}$ and, for $\emptyset \neq S \subset I$, $V(S)$ is a nonempty, closed cylinder set. Moreover, $V(S)$ is lower comprehensive, i.e. $V(S) + \mathbb{R}^I_- \subset V(S)$. The correspondence $V(S)$ assigns to each coalition S a set of utility allocations.

To define the cooperative game generated by an exchange economy with asymmetric information, the set of agents is identified with the set of players. The characteristic function is defined as follows:

$$V_k(\emptyset) := \{0\}$$

$$
\begin{aligned}
V_k(S) \;:=\; &\Big\{ z \in \mathbb{R}^I \,\big|\, \exists\, g = ((x_i, P_i))_{i \in I} \\
&\text{with } \sum_{i \in S} x_i = \sum_{i \in S} e_i \text{ where } P_i = k_S^i \left((P_j^0)_{j \in S} \right) \\
&\text{and } U_i(x_i, P_i) \geq z_i \; \forall i \in S \Big\},
\end{aligned}
$$

where z_i denotes the payoff to player i. The characteristic function associates with each nonempty coalition those payoff vectors $z = (z_i)_{i \in I}$ that are dominated by utility allocations of enforceable allocations. This definition imposes no restriction on the utility allocations of players outside the coalition.

The set F of feasible payoff allocations is defined as follows:

Definition 3.19 *The set F^k denotes the feasible payoff alloca-tions given the information rule k, i.e.*

$$F^k := \{z \in \mathbb{R}^I | \exists \left((x_i, P_i)\right)_{i \in I} \in E^k,$$
$$\text{with } U_i(x_i, P_i) \geq z_i \, \forall i \in I\}.$$

The set F^k contains all payoff vectors dominated by utility alloca-tions which can be generated by feasible allocations. The follow-ing theorem states that $V(S)$ and F have the required properties.

Theorem 3.1 *The sets $V(S)$ and F are nonempty, closed, and lower comprehensive cylinder sets. Moreover, $V(S) \cap \mathbb{R}^S$ and F are bounded from above.*

Proof. By definition, $V(\emptyset) = \{0\}$. Obviously, $V(S)$ and F are lower comprehensive. For $S \neq \emptyset$ the set $V(S)$ is nonempty, since the endowment is an allocation satisfying the inequality $U_i(e_i, P_i) \geq z_i$ for $z_i = U_i(e_i, P_i)$. Now $V(S) = \{V(S) \cap \mathbb{R}^S\} \times \mathbb{R}^{I \backslash S}$. Since $\mathbb{R}^{I \backslash S}$ is closed in $\mathbb{R}^I$, it suffices to show that the first factor is closed in $\mathbb{R}^I$ to proof that $V(S)$ is closed. Now $V(S) \cap \mathbb{R}^S = M + \mathbb{R}^S_-$, where

$$M := \{z \in \mathbb{R}^S | \exists \, g = \left((x_i, P_i)\right)_{i \in I}$$
$$\text{with } \sum_{i \in S} x_i = \sum_{i \in S} e_i, \text{ where } P_i = k^i_S \left((P^0_j)_{j \in S}\right)$$
$$\text{and } U_i(x_i, P_i) = z_i \forall i \in S\}.$$

Since $U_i(x_i, P_i) = U_i(x_i, \hat{P}_i)$ if $(x_i, P_i), (x_i, \hat{P}_i) \in X_i$, the set M can also be written as:

$$M := \{ (U_i(x_i, \Omega^*))_{i \in S} | \exists \, g = \left((x_i, P_i)\right)_{i \in I}$$
$$\text{with } \sum_{i \in S} x_i = \sum_{i \in S} e_i,$$
$$\text{where } P_i = k^i_S \left((P^0_j)_{j \in S}\right) \, \forall i \in S\}.$$

Being a continuous image of a compact set, the set M is itself compact. Its asymptotic cone $C(M)$ is $\{0\}$. Thus, $C(M)$ and $C(\mathbb{R}^S_-)$ are positive semiindependent. Therefore, $M + \mathbb{R}^S_-$ is closed. Since $\left(M + \mathbb{R}^S_-\right) \cap \mathbb{R}^S_+ = V(S) \cap \mathbb{R}^S_+$, and $\left(M + \mathbb{R}^S_-\right) \cap \mathbb{R}^S_+$ is bounded from above (because M is bounded from above), it follows that $V(S) \cap \mathbb{R}^S_+$ is bounded from above. A similar argument concerning the set F (replace $k^i_S\left((P_j)_{j \in S}\right)$ by P^m_i and S by I) shows that F is closed and bounded from above.

We have shown that an economy with asymmetric information generates a well defined cooperative game. □

Lemma 3.2 *The relation $V(I) \subset F$ holds.*

Proof. This follows immediately from the fact that

$$P^m_i \precsim k^i_I\left((P_j)_{j \in I}\right) \; \forall i \in I, \; k \in \mathcal{K}.$$

□

It can easily be seen that, in the case of the fine or private information rule, the set of feasible utility allocations coincides with the set of utility allocations the grand coalition can enforce: $F = V(I)$. In both cases, $P^m_i = k^i_I\left((P_j)_{j \in I}\right)$ holds.

We will now show that the core of any game generated by an exchange economy with asymmetric information is nonempty, regardless of the underlying information rule. A sufficient condition is that the game (I, F, V) be *balanced*.

Definition 3.20 *Let $\mathcal{B}$ be a family of nonempty subsets of I. The familiy $\mathcal{B}$ is called* balanced *if there exists for each $S \in \mathcal{B}$ a positive number λ_S such that*

$$\sum_{\substack{S \in \mathcal{B} \\ S \ni i}} \lambda_S = 1 \quad \forall i \in I.$$

Definition 3.21 *A game* (I, F, V) *is called balanced if, for all balanced families* $\mathcal{B}$,

$$\bigcap_{S \in \mathcal{B}} V(S) \subset F.$$

This definition of balancedness of a game differs from the one usually employed in the literature, where it is assumed that $F = V(I)$. As mentioned above, in the case of asymmetric information – as for example in the case of the coarse information rule – the possibility that a coalition can reallocate endowments only on the basis of an information which is coarser then the initial information of each member should not be excluded.

The proof that a balanced game (I, F, V) has a nonempty core is due to Böhm (1974).

Theorem 3.2 (Böhm 1974) *Let* (I, F, V) *be a balanced game. If F is nonempty, closed, bounded from above, and lower comprehensive, then the game* (I, F, V) *has a nonempty core.*

The following theorem states that any game generated by an exchange economy with asymmetric information is balanced and thus has a nonempty core, provided the agents' preferences are convex.

Theorem 3.3 *If the agents have quasiconcave utility functions, all games generated by exchange economies with asymmetric information are balanced.*

Proof. Let $z \in \cap_{S \in \mathcal{B}} V(S)$ and, for $S \in \mathcal{B}$, let $g^S = \left((x_i^S, P_i^S) \right)_{i \in I}$ be an allocation such that g^S is enforceable by S and $U_i(x_i^S, P_i^S) \geq z_i \ \forall \ i \in S$. Thus, $x_i^S - e_i$ is measurable with respect to $k_S^i \left((P_j^0)_{j \in S} \right)$, i.e. $x_i^S - e_i \in Map_{k_S^i((P_j^0)_{j \in S})}(\Omega, \mathbb{R}^L) \ \forall i \in S, \ \forall S \in \mathcal{B}$.

Since $P_i^m \precsim k_S^i \left((P_j^0)_{j \in S} \right)$ it follows that $x_i^S - e_i \in Map_{P_i^m}(\Omega, \mathbb{R}^L)$.
Now let $x_i = \sum_{\substack{S \in B \\ i \ni S}} \lambda_S x_i^S$. Each $x_i^S - e_i$ being measurable with
respect to P_i^m, it follows that $x_i - e_i$ is measurable with respect to
P_i^m, since convex combinations of P_i^m–measurable mappings are
P_i^m–measurable.

As for physical feasibility, the following holds:

$$\sum_{i \in I} x_i = \sum_{i \in I} \sum_{\substack{S \in B \\ S \ni i}} \lambda_S x_i^S = \sum_{S \in B} \lambda_S \sum_{i \in S} x_i^S =$$

$$\sum_{S \in B} \lambda_S \sum_{i \in S} e_i = \sum_{i \in I} \sum_{\substack{S \in B \\ S \ni i}} \lambda_S e_i = \sum_{i \in I} e_i.$$

This implies $\left((x_i, P_i^m) \right)_{i \in I} \in E$.

Since the utility functions are quasiconcave (preferences are convex), it follows that

$$U_i(x_i, P_i^m) = \tilde{U}_i(x_i) \geq z_i.$$

Thus, z_i is dominated by $U_i(x_i, P_i^m)$, and therefore z is an element
of F. Consequently, the game generated by an exchange economy
with asymmetric information is balanced. $\square$

The theorem provided by Böhm implies that the core of this game,
and therefore the core of the underlying economy, is nonempty.
Thus we have shown that an exchange economy in which the
agents might be asymmetrically informed about the states of nature possesses a nonempty core. This result is independent of the
information rule, i.e. the information the members of a coalition
can use for reallocating their endowments.

In general, the relationship between the different notions of the
core can be characterized as follows: The private information core
forms a subset of the coarse core: The set of feasible allocations is
the same for both information rules, but blocking an allocation is

'easier' under the private information rule, which involves weaker requirements with respect to measurability. However, the fine core will generally not be a subset of any other core. The reason is that the fine information rule typically generates a larger set of feasible allocations (as compared to other rules). Consequently, the fine core might comprise allocations that are not feasible under the coarse or private information rule.

3.4 Extensions: Incentives, Sequential Trade, and Learning

The model can be extended in several directions. First, the information rules have been introduced without discussing the problem of incentive compatibility.

3.4.1 Incentive Compatibility

An important question which has been raised by Koutsougeras and Yannelis (1993) and Krasa and Yannelis (1993) concerns the problem of incentive compatibility of a core allocation, i.e. whether there is truthful revelation of information in the core. To answer this question, Krasa and Yannelis have introduced the concept of coalitional incentive compatibility, which is defined as follows[6]:

Definition 3.22 *A feasible allocation* $f = ((x_i, P_i))_{i \in I}$ *is coalitionally incentive compatible for* $\mathcal{E}$ *if there is no coalition S and states* ω_k, ω_l *such that*

1. $P_i(\omega_k) \in k_S^i\left((P_j)_{j \in S}\right)$ *for all $i \in S$,*

2. $\omega_k \in P_i(\omega_l) \in k_{I \setminus S}^i\left((P_j)_{j \in I \setminus S}\right)$ *for all $i \in I \setminus S$,*

---

[6]Krasa and Yannelis call this weakly coalitional incentive compatible.

3. $\tilde{U}_i(e_i(\omega_k) + x_i(\omega_l) - e_i(\omega_l)) > \tilde{U}_i(x_i(\omega_k))$ for all $i \in S$.

In words: A feasible allocation is coalitionally incentive compatible if no coalition S can be formed such that the members of that coalition could 'cheat' about the realization of a state of nature with respect to agents in the coalition $I \setminus S$ who are unable to distinguish between these two states. Condition 1 says that all members in S should agree about the occurrence of the state ω_k, condition 2 states that all agents not in S cannot distinguish between ω_k and ω_l and the last condition states that the members of coalition S are better off by announcing a false state.

However, core allocations are not necessarily coalitionally incentive compatible. It has been shown by Koutsougeras and Yannelis (1993) that for the private and coarse information rule the core allocations are coalitionally incentive compatible, provided that the preferences are monotone. However, the fine core allocations are in general not coalitionally incentive compatible. This is illustrated by the following example taken from Koutsougeras and Yannelis (1993):

Example 3.4 Consider an economy with $\Omega = \{\omega_1, \omega_2, \omega_3\}$, one physical commodity and three agents with initial information

$$P_1^0 = \{\{\omega_1, \omega_2\}, \{\omega_3\}\},$$

$$P_2^0 = \{\{\omega_1, \omega_3\}, \{\omega_2\}\},$$

$$P_3^0 = \{\{\omega_1\}, \{\omega_2, \omega_3\}\}.$$

The endowments are given by

$$e_1 = (10, 10, 0, P_1^0),$$

$$e_2 = (10, 0, 10, P_2^0),$$

$$e_3 = (0, 0, 0, P_3^0).$$

The utility functions of the agents are given by $\tilde{U}_i = \sqrt{x_{i1} x_{i2} x_{i3}}$ for $i = 1, 2, 3$.

For the fine, coarse and private information rules, the following maximal information results: $P_i^{mf} = \Omega^*$, $P_i^{mc} = P_i^{mp} = P_i^0$. In words: Under the fine information rule there is no measurability constraint for feasible allocations. Under any of the other information rules, feasible allocations have to be measurable with respect to the agents' initial information. Of course, the initial endowments always represent a feasible allocation.

Now consider the feasible allocation

$$(x_1^*, P_1) = (8, 8, 2, P_1^0)$$

$$(x_2^*, P_2) = (8, 2, 8, P_2^0)$$

$$(x_3^*, P_3) = (4, 0, 0, P_3^0)$$

This allocation is a private information as well as a coarse core allocation: It is feasible and it is unblocked by any coalition given the private or coarse information rule. Notice that this allocation takes into account the superior information of agent 3: he ends up with a positive consumption despite the fact that he has only a zero endowment of contingent commodities. As Koutsougeras and Yannelis (1993) have shown, allocations in the private as well as in the coarse core are coalitionally incentive compatible.

However the allocation (x_i^*, P_i^0) is not an element of the fine core. The reason is that by pooling their information, agents 1 and 2 have perfect information. As a consequence the allocation (x_i^*, P_i^0) would be blocked by the coalition $\{1, 2\}$ using e.g. the allocation

$$(y_1^*, P_1) = (y_2^*, P_2) = (10, 5, 5, \Omega^*)$$

$$(y_3^*, P_3) = (0, 0, 0, P_3^0).$$

Notice that if state ω_1 occurs, the coalition $S = \{1\}$ will be better off by announcing state ω_3, since these two states cannot be distinguished by agent 2, given his initial information. However, the definition of coalitional incentive compatibility implies that *all* members of the coalition $I \setminus S$ should be unable to discriminate between states ω_1 and ω_3. Since agent 3 *is* able to distinguish between these two states and the fine information rule applies, the allocation is coalitionally incentive compatible. The reason is that, by forming a coalition with agent 3, agent 2 could use agent 3's superior information as an insurance against any cheating by agent 1. If however the initial information of agent 3 is changed to $P_3^0 = \{\{\omega_1, \omega_2, \omega_3\}\}$, the allocation is not coalitionally incentive compatible. This shows that fine core allocations are in general not coalitionally incentive compatible. The example suggests that superior information of an agent may affect incentive compatibility. His information can be used by other agents as an insurance device against cheating.

The private information core always generates allocations which are coalitionally incentive compatible. However, there are allocations which are incentive compatible but inconsistent with the private information core. It is still an open problem whether there is an information rule which generates all incentive compatible allocations as core allocations.

3.4.2 Sequential Trade and Learning

So far it has been assumed that transactions have to be agreed upon ex ante, i.e. before the state of the world is known, and are carried out after the state has been realized and is known by the agents. Stated otherwise, there is no additional trading after the state is known, and agents end up consuming what has been agreed upon prior to its realization.

However, if there are no institutions to enforce the fullfillment of the agreements or if there is no trust (Gale (1978)), it is not certain that every party of a transaction will stick to the contract agreed upon ex ante, i.e. there might be an additional trading round after the realization of the state. These problems have been discussed in the literature under the name of the sequential core (e.g. Gale (1978), Honkapohja (1977) and Repullo (1988)).

But even if such institutions exist or in the presence of trust, similar problems might arise in the case of asymmetric information, as pointed out by Koutsougeras (1995). If there are incentive compatibility problems ex ante, an additional round of trading after realization of the state will increase the cheating opportunities for coalitions and thus lead to an aggravation of the problem of incentive compatibility.

To analyze these incentive problems which arise if there are two rounds of trading, Koutsougeras (1995) introduced the concept of the two stage core. Before discussing this type of core in the framework considered here, some notions have to be redefined due to the additional trading after a state of the world is realized. In what follows, the first or ex ante stage, where agreements are made, is be denoted by 1, the second or ex post stage by 2. As in the model by Koutsougeras, it is assumed that the transactions agreed upon in stage 1 will be carried out and result in a new allocation. For each state of the world, the new commodity bundles are considered as the endowments for the ex post stage. Given these endowments, agents can form new coalitions and engage in further transactions. This implies that the coalitions which have formed in stage 1 will possibly be broken up since, with the new endowments, different coalitions could guarantee their respective members a higher utility.

Thus, the consumption set of agent i in the second period is given by

$$X_i^2 := \left\{ (x_i, P(\omega)) \in Map(\Omega, \mathbb{R}_+^L) \times \Omega^* | P(\omega) \in \Omega^* \right\}.$$

By abuse of notation, an element of X_i^2 will be denoted by $x_i(\omega)$.

Notice that, since there is no uncertainty in the second period, no measurability conditions with respect to information have to be satisfied.

The consumption set for the ex ante stage is given as

$$X_i^1 := \left\{ (x, P) \in Map(\Omega, \mathbb{R}_+^L) \times P^* | x - e_i \in Map_P(\Omega, \mathbb{R}^L) \right\} \cup X_i^2.$$

Notice that the information P used in the definition of X_i^1 depends on the information rule as well as the coalition agent i belongs to. A trading plan for agent i consists of a tuple $((x_i^1, P_i), x_i^2) \in X_i^1 \times X_i^2$. Since it is assumed that agents consume only after the second trading round when the state of the world is known, agent i's ex post utility is given by $\tilde{U}_i(x_i^2(\omega))$, while his ex ante utility is the same as defined in section 3.1.3, i.e. $U_i(x_i^1, P_i)$.

Assume that agents have agreed upon transactions leading to an allocation in state ω $((x_i'(\omega))_{i \in I}$ with $x_i'(\omega) \in X_i^2$ for all i. Taking this allocation as the new endowment, the ex post core can now be defined as follows:

Definition 3.23 *Given an allocation $((x_i', P_i))_{i \in I}$ with $(x_i', P_i) \in X_i^1$, the ex post core is the set of allocations $((x_i(\omega))_{i \in I}$ such that*

1. *$\sum_{i \in I} x_i(\omega) = \sum_{i \in I} x_i'(\omega)$ for each ω ,*

2. *there is no coalition $S \subset I$ and no allocation $((y_i(\omega))_{i \in I}$ with $y_i(\omega) \in X_i^2$ for all i such that $\sum_{i \in S} y_i(\omega) = \sum_{i \in S} x_i'(\omega)$ and $\tilde{U}_i(y_i(\omega)) > \tilde{U}_i(x_i'(\omega))$.*

In words: The ex post core consists pointwise for each ω of all allocations which are physically feasible and unblocked by any coalition.

Notice that the ex post core takes into account that the transactions an agent can make in period two depend on the arrangements agreed upon in period one.

With these notions the two stage core as introduced by Koutsougeras (1995) can be defined as follows:

Definition 3.24 *A trading plan* $((x_i^1, P_i), x_i^2) \in X_i^1 \times X_i^2$ *is in the two stage k-core if*

1. *the allocation* $((x_i^1, P_i))_{i \in I}$ *is k–feasible;*

2. $x^2 \in C^1(\mathcal{E}, x^1);$

3. *there exists no coalition which k–blocks the allocation* $((x_i^1, P_i))_{i \in I}.$

By arguments analogous to those in section 3.3 it can be shown that the two stage core is nonempty. In general however, the two stage core will in general not be coalitionally incentive compatible if the information rule allows for some pooling of information. Notice that in example 1 the fine core allocation is also in the two stage core but is not coalitionally incentive compatible if agent 3 is uninformed. Koutsougeras (1995) has shown that the private information two stage core is also coalitionally incentive compatible.

The following example illustrates the differences between the core as discussed in section 3.2.4 and the two stage core.

Example 3.5 Consider an economy with $\Omega = \{\omega_1, \omega_2, \omega_3\}$, two physical commodities and three agents with initial information

$$P_1^0 = \{\{\omega_1, \omega_2\}, \{\omega_3\}\},$$

$$P_2^0 = \{\{\omega_1, \omega_3\}, \{\omega_2\}\},$$

$$P_3^0 = \{\{\omega_1\}, \{\omega_2, \omega_3\}\}.$$

The endowments are given by

$$e_1 = ((20, 1), (20, 1), (0, 1), P_1^0),$$

$$e_2 = ((20,1),(0,1),(20,1),P_2^0),$$

$$e_3 = ((10,1),(10,1),(10,1),P_3^0).$$

The utility function of agent i is given by $\tilde{U}_i = \prod x_{ijk}$ for all i, $j = 1,2$; $k = 1,2,3$.

Given the fine information rule, the following allocation is in the fine core:

$$x_1 = ((20,1),(10,1),(10,1),P_1^f),$$

$$x_2 = ((20,1),(10,1),(10,1),P_2^f),$$

$$x_3 = ((10,1),(10,1),(10,1),P_3^f).$$

As in example 1, the coalition $S = \{1\}$ could announce ω_3 if the true state is ω_1, i.e. coalition S could cheat. Since agent 3 is able to discriminate between states ω_1 and ω_2 the definition 5.1 implies that the allocation $\left((x_i, P_i^f) \right)_{i \in I}$ is coalitionally incentive compatible. Recall that incentive compatibility requires that *all* agents not in the coalition are unable to discriminate between these two states. However, the allocation is not in the two stage fine core, since in state ω_1 agent 1 ends up with $x_1(\omega_1) = (30,1)$ by announcing state ω_3. Here, agent 1 keeps his endowment and receives 10 units of the commodity from agent 2. Forming a coalition with agent 3, the coalition $S = \{1,3\}$ could achieve a higher utility by a transaction leading e.g. to $x_1(\omega_1) = (27,1.2)$ and $x_3 = (13,0.8)$. While the ex ante transactions lead to utilities of 30 and 10 for agents 1 and 3 in state ω_1, ex post transactions yield utilities of 32.4 and 10.4 for the two agents.

This shows that the possibility of ex post trading has an important impact: Allocations which are in the fine core and are also coalitionally incentive compatible are in general not in the two stage fine core.

The problem of sequential trade is closely related to the question whether agents are able to learn over time. By observing realized allocations or transactions agents could deduce information about the states of nature. Such an approach would allow to dispense with exogenously given information rules and to endogenize the change of information. This problem has been addressed in the paper by Koutsougeras and Yannelis (1995).

They consider a model of an exchange economy similar to the one discussed in section 3.1 but which extends over time. Concentrating on the private information rule and the corresponding core concept, Koutsougeras and Yannelis (1995) consider the following learning process: In each period, each agent is able to observe the sigma algebra generated by his initial information, his endowment, his utility function and all allocations realized in previous periods. Even without information sharing this learning process will generate successively finer information for each agent. Koutsougeras and Yannelis (1995) show that this process will converge to full or perfect information for each agent and thus to the full information coreif this process is repeated infinitely often. Stated otherwise, in the limit asymmetric information will disappear since each agent is perfectly informed about the states of nature. This implies that in the limit there are no incentive compatibility problems.

Notice that the full information which each agent has gained in the limit differs from the maximum amount of information as defined in section 3.2.3. While the former corresponds to $P_i = \Omega^*$, the maximum amount of information is generally different from perfect or full information. If there is no information sharing, i.e. if the private information rule applies, the maximum amount of information is equal to an agent's initial information. Under the fine information rule, however, both concepts lead to identical results.

However, the assumption made by Koutsougeras and Yannelis (1995) that allocations are observed is not directly applicable to our model with information contingent commodities. In this context, the assumption of observable allocations would imply that the agents be able to observe each other's information directly. It follows that the fine core, or the full information core, would result even with no learning over time.[7]

This gives rise to the question whether the assumption of an allocation being observed might be too restrictive for the model discussed here. Alternatively, one could assume observability of transactions of physical commodities only, i.e. observations are made with respect to the economic activities of agents. In the present context, the latter assumption is perhaps of greater economic interest. The following example taken form Koutsougeras and Yannelis (1995) shows that these assumptions imply different results.

Example 3.6 There is only one commodity, three states of nature and two agents 1 and 2. Their initial information is given by

$$P_1^0 = \{\{\omega_1, \omega_2\}\{\omega_3\}\}, \quad P_2^0 = \{\{\omega_1, \omega_3\}\{\omega_2\}\},$$

and their endowments are

$$e_1 = (10, 10, 0, P_1^0), \quad e_2 = (10, 0, 10, P_2^0).$$

Notice that if there is no information sharing, only the vector of initial endowments is in the core. However, if agents are able to observe allocations, their information in the next period will be perfect information.[8] In this case, the private information core allocation is

$$x_1 = (10, 5, 5, \Omega^*), \quad x_2 = (10, 5, 5, \Omega^*).$$

---

[7]In the present context, an assumption equivalent to that of Koutsougeras and Yannelis (1995) would be that agents can observe only the state contingent part of an allocation, but not the information itself.

[8]One could ask the question why Koutsougeras and Yannelis (1995) assume that agents don't observe the initial endowments and arrive at the full information core immediately since a vector of initial endowments is an allocation as any other.

However, if only transactions are observed, the result is different. Without information sharing, no transactions can be carried out, i.e. each agent consumes his initial endowment. Since there is no economic activity, no information is generated. Thus, the process will not converge to the full information core. In each period, the agents' information will not change, and the only core allocation is the vector of initial endowments.

This notion of learning by observing transactions establishes an interesting relationship between models of the core with asymmetric information and models of general equilibrium with rational expectations as e.g. developed by Radner (1979). In these models, agents might be able to learn from the equilibrium prices about the other agents' non–price information. However, if there is no economic activity, equilibrium prices will not change and thus there is no opportunity for the agents to learn. As a consequence, the equilibrium prices in the rational expectations model play a role similar to the transactions that lead to a core allocation in the cooperative approach.

3.5   Conclusion

This chapter studies the core of exchange economies with asymmetric information. Each agent's information is modelled as a partition of a finite set of states of nature. In contrast to the existing literature, information forms a constituent part of the economy. The agents' accessible information may change by their joining a coalition. The process of information transmission is described by an exogenously given information rule as introduced by Allen (1991). An information rule defines the information the members of a coalition can use to reallocate their endowments.

The set of core allocations varies with the underlying information rule. In contrast to the results derived by Allen and Yannelis, we show that the core of an exchange economy with asymmetric information is always nonempty, irrespective of the information rule. This discrepancy is mainly due to the different notions of feasibility employed in the respective models. While Allen considers as feasible only those allocations that can be achieved by net trades compatible with the information accessible by the members of the grand coalition, Yannelis demands that feasible allocations be compatible with each coalition member's initial information. In our model, an allocation is feasible if it can be achieved by net trades compatible with the maximum amount of information in the economy, i.e. the information each agent could gain if he joined all possible coalitions simultaneously. Our feasibility definition is thus independent of any particular coalition structure.

Apart from the extensions mentioned in section 3.4, there is a variety of interesting questions that can be addressed in the framework of cooperative models with asymmetric information.

First, extending the analysis to coalition production economies might contribute new ideas to the theory of the firm in general equilibrium models. In this context, the recent book by Ichiishi on 'The Cooperative Nature of the Firm' deserves mentioning. However, he excludes problems of asymmetric information. This issue will be addressed in the next chapter. Secondly, the commodity space as introduced here allows for the examination of Walras equilibria in economies with asymmetric information. This approach might be used to explain a widely observed phenomenon, namely that, in equilibrium, agents with different information pay different prices for the same physical or state–contingent commodity.

Chapter 4

Production Economies

Except for some remarks in the paper by Wilson (1978), the literature on the core of economies with asymmetric information has focused on the analysis of pure exchange economies. Up until now, the core of a production economy with asymmetric information has not been analyzed in the literature.

This is a major shortcoming since asymmetric information may have a considerable impact on employment and production, as a large number of results in the theory of optimal contracts, principal and agent, and industrial organization show. Asymmetric or differential information will generally lead to inefficient employment and an inefficient level of production, as compared to the complete information case. However, these results have been derived in a partial equilibrium framework and it is not clear whether they are also valid in a general equilibrium framework. Therefore, the approach developed in chapter 3 will be extended to the analysis of production economies.

The chapter is organized as follows: In the first section a model of a production economy with asymmetric information is developed. Production is modelled by associating to each coalition a technology set, i.e. a set of technologically feasible production plans. Such an economy is known in the literature as a coalition production economy. However, the production plans a coalition can carry out depend not only on the available technology but also on the information the coalition can employ when choosing a production plan, which in turn depends on the information of the coalition members.

In section 4.2 the set of feasible allocations in a coalition production economy is discussed. As before, the feasible set is determined also by the maximum amount of information that can be used in production by the members of the coalition. Also the mathematical concept of an asymptotic cone of a set is extended to the commodity space considered here.

The following sections deal with the core of coalition production economies and with the games generated by these economies. It can be shown that, if the technology sets satisfy some common regularity conditions, a production economy with asymmetric information generates a cooperative game. Further, the core of this game, and thus the core of the underlying economy, is nonempty, provided that the technology sets satisfy a balancedness condition. This balancedness condition has been developed by Böhm (1974a, 1974b) and has proved to be of vital importance for the existence of core allocations. It turns out that the condition of balancedness – accordingly modified for the case of asymmetric information – is also of fundamental importance for the type of economies considered here. The result that the core of the economy is nonempty holds independently of the information that agents and coalitions can employ with respect to transactions and for production. The impact of asymmetric information on an economy with production is illustrated by a simple example. While it has been assumed that the set of enforceable allocations for each coalition is nonempty, the next section allows for the coalition formation to be costly. This might lead to the case that, if these costs are sufficiently high, a coalition with more than one member can enforce no allocation whatsoever. However, it can be shown that even if there are costs of coalition formation, the core of the economy is nonempty. The chapter closes with a summary and a discussion of the results.

4.1 Coalition Production Economies with Asymmetric Information

In this section, the model developed in chapter 3 will be extended by allowing for production. The members of a coalition use their endowments not only for a reallocation of resources, but also for production. For this purpose, each coalition is assigned a technology set and the information it can employ in production. Thus, the possible production decisions of a coalition, or 'firm', are constrained by the available technology and by the information the coalition can employ in production. Besides these modifications, all other assumptions concerning the states of nature, the information of the agents, the commodity space, consumption sets, and initial endowments remain unchanged.

4.1.1 Production Information Rules

To determine a coalition's production possibilities, it is necessary to define the information a coalition can employ when choosing a production plan. In addition, the technology or production set of a coalition has to be specified.

To characterize the information a coalition can employ in production, a similar approach is chosen as the one that describes the information the members of a coalition can use for the reallocation of their endowments. In the case considered here, the information a coalition can employ in production is also determined by an information rule.

This mapping, called the *production-information-rule*, is defined as follows:

Definition 4.1 *A* production information rule *for a coalition $S \subset I$ is a mapping*

$$t_S : (P^*)^S \to P^*$$

with the property $t_S = Id$ if $|S| = 1$, i.e. singleton–coalitions and agents are identified. A production-information-rule for the economy is a $2^{|I|}$–tuple $t = (t_S)_{S \in \mathcal{P}(I)}$ of production-information-rules, one for each coalition.

On a conceptual level, one should distinguish between the information a coalition can use for production and the information the members of the coalition may employ to reallocate their endowments. That is, the information with which coalition S is associated by the production information rule is a priori unrelated to the information the members of the coalition receive by the rule k_S. This conceptual discrimination between the two kinds of information is in accordance with the double role the agents have in this economy: first as consumers and second as producers. This double role of an agent in a coalition production economy has been pointed out by Ichiishi in his work on the cooperative nature of the firm: 'Each agent plays at least two roles in the economy: that of a consumer, and that of a member of the firm he works for.'[1]

To distinguish between the information which can be used in production and the information which the members of the coalition attain by the information rule k_S, the former is denoted by Q_S. In addition, the information rule t_S differs from the rule k_S in that t_S associates *one* information to a *coalition as a whole*, while the rule k_S associates possibly a different information to *each member* of the coalition. The reason for this assumption is that the choice of a production plan is considered a *joint* or *collective decision* of the coalition based upon a *single* information. For example, one could imagine that one member of the coalition is in charge of choosing a production plan. In this case, the information of this member is relevant to the production decision. This agent can be considered as the 'manager' of the coalition. Alternatively, the information rule t_S could also describe the case in which the members of the coalition aggregate their information and choose a production plan compatible with this information.

---

[1]Ichiishi (1993), p.79.

As an information Q is associated with each coalition in its role as producer, the notions of *perfect* and *symmetric* information have to be modified accordingly. Here, the information structure of an economy is given by $((P_i)_{i \in I}, (Q_S)_{S \subset I})$.

Definition 4.2 *An information P_i of an agent i or an information Q_S of a coalition S is called* complete *if $P_i = \Omega^*$ or $Q_S = \Omega^*$. An information structure of the economy $((P_i)_{i \in I}, (Q_S)_{S \subset I})$ is called* complete *if the information of each agent and each coalition is complete. Otherwise, an information is called* incomplete.

Definition 4.3 *An information structure of the economy $((P_i)_{i \in I}, (Q_S)_{S \subset I})$ is called* asymmetric *if there are at least two agents i, i' with $P_i \neq P_{i'}$ or two coalitions S, S' with $Q_S \neq Q_{S'}$.*

Having introduced the concept of the information a coalition can employ in production, the technology set of a coalition has to be defined.

4.1.2   The Technology Set of a Coalition

Before describing the technology set of a coalition, some useful notation is introduced.

Definition 4.4 *The set $Map(\Omega, \mathbb{R}^L) * P^* \subset Map(\Omega, \mathbb{R}^L) \times P^*$ consists of pairs (y, Q), such that $y \in Map_Q(\Omega, \mathbb{R}^L)$.*

Form the set $Map(\Omega, \mathbb{R}^L) \times P^*$, the subset $Map(\Omega, \mathbb{R}^L) * P^*$ is distinguished. It contains all pairs (y, Q) such that y is compatible with Q.

The technology set for each coalition S is given by a nonempty set $Y(S) \subset Map(\Omega, \mathbb{R}^L) * P^*$. The elements of such a technology set, the production plans, are ordered pairs (y, Q) such that y is compatible with the information Q. For the empty coalition, the relation $Y(\emptyset) = \{0\} \times \{\Omega\}$ holds. Thus, by definition production plans can not be considered independently of the information. State contingent inputs and outputs of the production process can be chosen only contingent on events which are compatible with the information the coalition can employ in production. Such a technology set $Y(S)$ describes all net output bundles that a coalition S can make available to its members through a *joint* action.

In the following, it is assumed that a coalition is always able to dispense with production. In section 4.6 the analysis is extended to the case where the formation of a coalition is costly which implies that resources have to be employed to form a coalition. For such a coalition a production plan without inputs is not feasible. Moreover, as Böhm (1974a) has pointed out, the production decisions of two coalitions S_1 and S_2, i.e. $(y_1, Q_{S_1}) \in Y(S_1)$ and $(y_2, Q_{S_2}) \in Y(S_2)$, are not necessarily related to the production decision of the coalition $S = S_1 \cup S_2$.[2] In the context considered here, the information of the coalition $S_1 \cup S_2$ could be coarser than the information the subcoalitions can employ in production. In this case, production plans which are feasible for the subcoalitions might be infeasible for the grand coalition due to less information. But the opposite case is also possible. As Ichiishi pointed out: 'To provide a powerful explanation of formation of firms, one needs to allow for the situation in which a coalition has a more efficient production set than the sum of the production sets held by the individuals of the coalition, that is, the production–set correspondence is non–additive.'[3]

In accordance with Böhm (1974a), a total production possibility set $Y \subset Map(\Omega, \mathbb{R}^L) * P^*$ is defined. This production possibility set is not necessarily identical to the technology set of the grand coalition. For example, if there are coordination problems that become more and more severe the larger the coalition gets, production could be more efficient for smaller coalitions. Thus, it is in general not true that all feasible production plans can be en-

---

[2]Böhm (1974a), p.430
[3]Ichiishi (1993) p.116

forced by the grand coalition. In other words, it is not assumed that $Y = Y(I)$. As Böhm writes: 'In this case there would exist a true incentive for decentralization which may result in a Pareto superior allocation to any allocation which is enforceable by the grand coalition I.'[4]

The technology sets available in the economy can now be described by a family $T = ((Y(S))_{S \subset I}, Y)$ of nonempty subsets of $Map(\Omega, \mathbb{R}^L) * P^*$. This family will be denoted as the *technology of the economy*. Using the concepts developed in chapter 3, the production information rule and the technology of the economy, a *coalition production economy with information rules*, $\mathcal{E}^{kt}$, can be defined as follows:

$$\mathcal{E}^{kt} := \left(I, \left(X_i, U_i, e_i, P_i^0 \right)_{i=1}^{|I|}, T = ((Y(S))_{S \subset I}, Y), k, t \right).$$

Apart from the technology of an economy and the production information rule, this definition corresponds to the definition of an exchange economy with information rule. For the case $Y(S) = \{(0, Q)\}$ with $Q \in P^*$ for all $S \subset I$ and $Y = \{(0, Q)\}$ with $Q \in P^*$, i.e. if there are no production possibilities, the economy is equivalent to an exchange economy with information rule. If there is only one state of nature, the definition describes a coalition production economy in the usual sense.[5]

4.2 Feasible Allocations

As in an exchange economy with information rule, an *allocation* is defined as a tuple $((x_i, P_i)_{i \in I})$, where $(x_i, P_i) \in X_i$ for all $i \in I$.

To define the set of feasible allocations for this economy, the maximum amount of information has to be determined. This is done

---

[4]Böhm (1974a), p.430. In section 4.4 an example is provided which shows that asymmetric information may give an incentive for decentralization.

[5]As defined e.g. in Böhm (1974a).

in a way analogous to that of an exchange economy. As there are two kinds of information in a coalition production economy, i.e. the information that can be used for exchange and for production, the maximum amount of information for each of these has to be determined.

The maximum amount of information of an agent i which can be employed for reallocation is, as in an exchange economy,

$$P_i^m := \bigvee_{S \ni i} k_S^i((P_i^0)_{i \in S}).$$

The maximum amount of information which can be used in production is defined by

$$Q^m := \bigvee_{\emptyset \neq S \subset I} t_S((P_i^0)_{i \in S}).$$

This information is the coarsest common refinement of all information which can be used in production. This can be interpreted as follows: If an information Q is considered the information available to a manager of a coalition, then the maximum amount of information in the economy is the information of the best informed manager. If there are several managers with incomparable 'best information', it is assumed these managers are able to pool their information.

Thus, the maximum amount of information contained in the economy can be written as $((P_i^m)_{i \in I}, Q^m)$. Notice that this concept is used only to describe the set of allocations which can at best be conceived of as feasible for the economy. It is not assumed that any agent or coalition has indeed access to this maximum amount of information.

Using this concept, the set of *feasible allocations* can now be defined as follows:

Definition 4.5 *An allocation $((x_i, P_i)_{i \in I})$ is called kt–feasible if there is a production plan $(y, Q) \in Y$ with*

1. $\sum_{i\in I} x_i = \sum_{i\in I} e_i + y$,

2. $x_i - e_i \in Map_{P_i^m}(\Omega, \mathbb{R}^L)$,

3. $y \in Map_{Q^m}(\Omega, \mathbb{R}^L)$.

These conditions imply that a feasible allocation has to be, first, physically feasible, and second, can be attained by net trades that are compatible with the maximal information P_i^m of the agents. Condition 3 guarantees that the production plan y is compatible with the maximum amount of information Q^m which can be used in production. The set of feasible allocations will be denoted by E^{kt}. If there is no danger of confusion, the set of feasible allocations will be denoted by E.

As it has been shown in the chapter on exchange economies, some properties of the set of feasible allocations – nonemptiness and compactness – are of central importance to prove the existence of core allocations.

To ensure that the set of feasible allocations satisfies these requirements, some assumptions concerning the technology are necessary. They are provided by the following lemmas.

Lemma 4.1 *If* $(0, Q) \in Y$ *for a* $Q \in P^*$, *than* $E \neq \emptyset$.

Proof. This follows immediately since the allocation $((e_i, P_i^0))_{i\in I}$ is feasible. $\qquad\square$

This lemma shows that if the general technology set allows for inaction, the set of feasible allocations is nonempty.

In economies with complete information, an assumption on the asymptotic cone of the technology set Y, denoted by $A(Y)$, guarantees the compactness of the set of feasible allocations. This approach can also be used in economies with asymmetric information. However, it has to be taken into account that the commodity space given here is not a Euclidean space. Therefore, the concept of an asymptotic cone has to be extended to the space considered here.

4.2.1   Asymptotic Cones in $Map(\Omega, \mathbb{R}^L) \times P^*$

To define the asymptotic cone of subsets of $Map(\Omega, \mathbb{R}^L) \times P^*$, the following notation is introduced: If M is a subset of $Map(\Omega, \mathbb{R}^L) \times P^*$ and M' a subset of $Map(\Omega, \mathbb{R}^L)$, the sets M_+ and M'_+ denote the 'non–negative' elements of M and M' respectively, i.e.

$$M_+ := M \cap \left(Map(\Omega, \mathbb{R}^L_+) \times P^* \right)$$

and

$$M'_+ := M' \cap Map(\Omega, \mathbb{R}^L_+).$$

If M is a subset of $Map(\Omega, \mathbb{R}^L) \times P^*$, than M_Q denotes the set

$$M_Q := \pi \left[M \cap \left(Map(\Omega, \mathbb{R}^L) \times \{Q\} \right) \right],$$

where π is the projection on the first component of $Map(\Omega, \mathbb{R}^L) \times \{Q\}$. A subset $Map(\Omega, \mathbb{R}^L) \times \{Q\}$ of $Map(\Omega, \mathbb{R}^L) \times P^*$ will be called a *'sheet'* of $Map(\Omega, \mathbb{R}^L) \times P^*$. The set M_Q is therefore a subset of the Euclidean space $Map(\Omega, \mathbb{R}^L)$. Obviously the following equality holds:

$$M \cap \left(Map(\Omega, \mathbb{R}^L) \times \{Q\} \right) = M_Q \times \{Q\}.$$

In what follows the set $M_Q \times \{Q\}$ is called a *sheet of M*. Furthermore,

$$M = \bigcup_{Q \in P^*} (M_Q \times \{Q\}).$$

The set M is the union of its sheets.

Using the set M_Q as a subset of the Euclidean space, the asymptotic cone of a set $M \subset Map(\Omega, \mathbb{R}^L) \times P^*$ can now be defined.

Definition 4.6 *If $M \subset Map(\Omega, \mathbb{R}^L) \times P^*$, then the set*

$$\tilde{A}(M) := \bigcup_{Q \in P^*} (A(M_Q) \times \{Q\})$$

is called the asymptotic cone *of M. Here $A(M_Q)$ denotes the asymptotic cone in the usual sense, since M_Q is a subset of the Euclidean space $Map(\Omega, \mathbb{R}^L)$.*

Because of the relationship

$$\tilde{A}(M \times \{Q\}) = A(M_Q) \times \{Q\} \cong A(M_Q)$$

$\tilde{A}$ can be considered as the extension of the concept of the asymptotic cone to subsets of $Map(\Omega, \mathbb{R}^L) \times P^*$. For this reason the asymptotic cone of a subset $M \subset Map(\Omega, \mathbb{R}^L) \times P^*$ will be denoted by $A(M)$.

The asymptotic cone of a set M consists – for each sheet – of the usual asymptotic cones of M_Q.

The set of 'non–negative' elements of the asymptotic cone $A(M)$ of M is of the form

$$A(M)_+ = \bigcup_{Q \in P^*} (A(M_Q)_+ \times \{Q\}) \,.$$

In particular, the statement

$$A(M)_+ = \{0\} \times P^*$$

is equivalent to

$$A(M_Q)_+ = \{0\} \text{ for all } Q \in P^*,$$

since the relationship

$$A(M)_+ = \bigcup_{Q \in P^*} (A(M_Q)_+ \times \{Q\})$$

obviously holds.

By using these notation, an important relationship between the closedness of a technology set Y as a subset of the commodity space $Map(\Omega, \mathbb{R}^L) \times P^*$, the asymptotic cone of this technology set, and the compactness of the set of non–negative state contingent production plans can be established.

Lemma 4.2 *If Y is closed in $Map(\Omega, \mathbb{R}^L) \times P^*$ and if $A(Y)_+ = \{0\} \times P^*$, then the set*

$$(Y_Q + v)_+$$

is compact for all $v \in Map(\Omega, \mathbb{R}^L)$.

Proof. The sets Y_Q are closed in $Map(\Omega, \mathbb{R}^L)$, and the statement $A(Y)_+ = \{0\} \times P^*$ implies $A(Y_Q)_+ = \{0\}$ for all $Q \in P^*$. Therefore $(Y_Q + v)_+$ is a compact subset of $Map(\Omega, \mathbb{R}^L_+)$ for all $v \in Map(\Omega, \mathbb{R}^L)$. $\square$

The lemma implies that the set of non–negative state contingent production plans of Y_Q is compact, i.e. closed and bounded. This holds true even if this set is shifted by a vector.

4.2.2   Properties of the Set of Feasible Allocations

The further analysis of the properties of the set of feasible allocations requires the introduction of some additional notation.

Let h denote the homeomorphism

$$\prod_{i \in I} \left(Map(\Omega, \mathbb{R}^L) \times P^* \right) \cong \prod_{i \in I} Map(\Omega, \mathbb{R}^L) \times \prod_{i \in I} P^*,$$

i.e.

$$h\left((x_1, P_1), (x_2, P_2), \ldots, (x_I, P_I) \right) =$$
$$(x_1, x_2, \ldots, x_I, P_1, P_2, \ldots, P_I) \quad .$$

Using the projection π on the first component of the product $\prod_{i \in I} Map(\Omega, \mathbb{R}^L) \times \prod_{i \in I} P^*$, the state contingent commodities can be separated from the information. That is

$$\pi(x_1, x_2, \ldots, x_I, P_1, P_2, \ldots, P_I) = (x_1, x_2, \ldots, x_I).$$

If $(P_i)_{i \in I} \in \prod_{i \in I} P^*$ is a tuple of agents' information in the economy, let $E_{(P_i)_{i \in I}}$ denote the set

$$E_{(P_i)_{i \in I}} := \left\{ (x_i)_{i \in I} \in \prod_{i \in I} Map(\Omega, \mathbb{R}^L) \mid (x_i, P_i)_{i \in I} \in E \right\}.$$

In particular, the set $E_{(P_i)_{i \in I}}$ is a subset of the Euclidean space $\prod_{i \in I} Map(\Omega, \mathbb{R}^L)$.

It is obviously true that

$$h(E) = \bigcup_{(P_i)_{i \in I} \in \prod_{i \in I} P^*} \left(E_{(P_i)_{i \in I}} \times \{(P_i)_{i \in I}\} \right).$$

If it can be shown that the set $E_{(P_i)_{i \in I}}$ is compact for all $(P_i)_{i \in I}$, the compactness of $h(E)$ follows immediately and also the compactness of E since h is a homeomorphism.

Theorem 4.1 *If Y is closed in $Map(\Omega, \mathbb{R}^L) \times P^*$ and if $A(Y)_+ = \{0\} \times P^*$, then the sets $E_{(P_i)_{i \in I}}$ are compact and therefore the set of feasible allocations E is compact.*

Proof. It is sufficient to show that $E_{(P_i)_{i \in I}}$ is compact for all $(P_i)_{i \in I}$. Let $(P_i)_{i \in I}$ be an element of $\prod_{i \in I} P^*$. The set $E_{(P_i)_{i \in I}}$ is thus a closed subset of

$$\{(x_i)_{i \in I} \mid \sum_{i \in I} x_i = \sum_{i \in I} e_i + y, \text{ where } y \in Y_Q \text{ for a } Q \in P^*\}.$$

Lemma 4.2 implies that the set $\left(Y_Q + \sum_{i \in I} e_i\right)_+$ is compact. Thus, there is a $w^Q \in Map(\Omega, \mathbb{R}^L)$ with $Y_Q + \sum_{i \in I} e_i \leq w^Q$. This implies

$$y + \sum_{i \in I} e_i \leq w^Q \leq \max_{Q \in P^*} \{w^Q\} \text{ for all } y \in Y_Q.$$

If $(x_i)_{i \in I}$ is an element of $E_{(P_i)_{i \in I}}$, the relationship $y + \sum_{i \in I} e_i = \sum_{i \in I} x_i \geq 0$ implies that $y + \sum_{i \in I} e_i \in \left(Y_Q + \sum_{i \in I} e_i\right)_+$. Together, this implies

$$0 \leq x_i \leq \sum_{i \in I} x_i = y + \sum_{i \in I} e_i \leq \max_{Q \in P^*} \{w^Q\},$$

and thus the compactness of $E_{(P_i)_{i \in I}}$. Therefore, also the set of feasible allocations is compact.   □

In a way similar to Böhm (1974a) it can be shown that the set of feasible allocations is compact if the asymptotic cone of the set Y satisfies certain conditions. Since a different commodity space is considered, the concept of an asymptotic cone would have to be modified accordingly.

4.3 The Core of a Production Economy with Asymmetric Information

As in the analysis of exchange economies, the concept of *blocking an allocation* by a coalition has to be defined. With production, one has to take into account that for blocking an allocation, a coalition can not only reallocate the endowments of its members but can also use these endowments for production. Notice, however, that these production plans have to be compatible with the information assigned to the coalition by the information rule t. Thus, *blocking an allocation* is defined as follows:

Definition 4.7 *A nonempty coalition $S \subset I$ blocks an allocation*

$$f = (x_i)_{i \in I}$$

if there is an allocation

$$g = (x_i')_{i \in I}$$

and a production plan $y' \in Y(S)$ with

1. $\sum_{i \in S} x_i' = \sum_{i \in S} e_i + y'$,

2. $x_i' - e_i$ *is measurable with respect to $k_S^i((P_i^0)_{i \in S})$,*

3. y' *is measurable with respect to $t_S((P_i^0)_{i \in S})$,*

4. $U_i(x_i') > U_i(x_i)$ *for all $i \in S$.*

Here, condition 1 implies that the allocation which is used for blocking has to be physically feasible for the coalition. The second condition states that the net trades of agent i are compatible with the information generated by the information rule k_S, while condition 3 requires that the production plan is in accordance with the information the coalition can employ in production. The last condition implies that the allocation g must give a higher utility to each member of the coalition.

Notice that this definition of blocking implies a relationship between the information the agents have at their disposal as consumers and the information that can be employed in production. Since $y_i = x_i' - e_i$ holds and $x_i' - e_i$ has to be compatible with the information P_i', it follows that y_i must also be compatible with P_i'. Thus, if the information for production is rather fine, e.g. $Q_S = \Omega^*$, but the information P_i' is coarse, e.g. $P_i' = \{\Omega\}$ for all $i \in I$, then the fine information Q_S cannot be used. Since in this case $x_i' - e_i$ is measurable only with respect to $\{\Omega\}$, it follows that $\sum_{i \in S} x_i' - \sum_{i \in S} e_i = y'$ is measurable with respect to $\{\Omega\}$. In other words: in spite of the fine information which could be employed in production $Q_S = \Omega^*$, the coalition can use only those production plans which are state independent for blocking. Such a situation could arise if, for example, in a coalition with two members one agent has complete information and acts as the 'manager' of the coalition, while the other has no information at all and the agents have no means of exchanging information truthfully so that only state independent transactions can be carried out.

The following definition describes the set of allocations a coalition can enforce.

Definition 4.8 *An allocation which satisfies conditions 1-3 in theorem 4.7 is called* enforceable by S. *The set of allocations enforceable by S is denoted by $E(S)$. If a specific technology $T = ((Y(S))_{S \subset I}, Y)$ is important for the analysis, the set of allocations enforceable by coalition S are denoted by $E_T(S)$.*

Obviously, $E(S)$ is a closed subset of $\prod_{i \in I} X_i$ if $Y(S)$ is closed in $Map(\Omega, \mathbb{R}^L) \times P^*$. The properties of the set of enforceable allocations $E(S)$ can be derived in a manner similar to the one used for deriving the properties of the set of feasible allocations E.

Lemma 4.3 *Is* $(0, Q) \in Y(S)$ *for a* $Q \in P^*$, *then* $E(S) \neq \emptyset$.

Proof. This follows immediately since $(e_i, P_i^0)_{i \in I}$ is enforceable. $\square$

In words: If a coalition's technology set allows for inaction, the set of enforceable allocations is nonempty since the vector of initial endowments is always an enforceable allocation.

Theorem 4.2 *If*

1. *$Y(S)$ is closed in* $Map(\Omega, \mathbb{R}^L) \times P^*$ *and*

2. *$A(Y(S))_+ = \{0\} \times P^*$,*

the set $\pi_S(E(S))$ *is compact. Here,* $\pi_S : \prod_{i \in I} X_i \to \prod_{i \in S} X_i$ *denotes the projection on the components* $i \in S$.

Proof. The proof is similar to the proof of theorem 1. $\square$

Since the set of allocations enforceable by coalition S contains no information about agents not in S, compactness of $E(S)$ can be shown only with respect to the appropriate subspace.

Now we can easily state under what conditions an allocation is an element of the core of an economy.

Definition 4.9 *An allocation* $f = ((x_i, P_i))_{i \in I}$ *is an element of the core of an economy* $\mathcal{E}$ *if the following conditions are satisfied:*

1. *the allocation f is feasible,*

2. *the allocation f is unblocked by any coalition.*

In what follows, the core of an economy $\mathcal{E}$ will be denoted by $Core(\mathcal{E})$. In order to analyze whether the core of a production economy is nonempty we have to make sure that such an economy generates a well–defined cooperative game. This is done in the next section.

4.4   The Cooperative Game Generated by a Production Economy

In general, a cooperative game (without transferable utility) is completely characterized by the set of players and the characteristic function. In our framework, however, the set of feasible utility allocations has to be taken into account. The reason is that, in our model, feasibility is not determined by what the grand coalition can achieve.

A cooperative game without transferable utility in characteristic function form is a tuple (I, F, V) where I denotes the set of players, F is the set of feasible payoff allocations, and V is a correspondence $V : \mathcal{P}(I) \to \mathbb{R}^I$ with the following properties: $V(\emptyset) = \{0\}$ and, for $\emptyset \neq S \subset I$, $V(S)$ is a nonempty, closed cylinder set. Moreover, $V(S)$ is lower comprehensive, i.e. $V(S) + \mathbb{R}^I_- \subset V(S)$. The correspondence V assigns a set of payoff allocations to each coalition S.

To define the cooperative game generated by an exchange economy with asymmetric information, the set of agents is identified with the set of players. The characteristic function is defined as follows:

For the empty coalition, the following holds:

$$V(\emptyset) := \{0\}.$$

For $S \neq \emptyset$,

$$V(S) := \{z \in \mathbb{R}^I \mid \exists \, g = ((x_i, P_i))_{i \in I} \in E(S)$$
$$\text{with } U_i(x_i, P_i) \geq z_i \,\forall\, i \in S\}.$$

A set $V(S)$ contains all those payoff allocations which are dominated by utility allocations generated by *enforceable* allocations.

Besides the payoff allocations enforceable by a coalition S, we have to define the set F of feasible payoff allocations.

$$F := \{z \in \mathbb{R}^I \mid \exists\, g = ((x_i, P_i))_{i \in I} \in E$$
$$\text{with } U_i(x_i, P_i) \geq z_i \,\forall\, i \in I\}.$$

The set F comprises all payoff allocations which are dominated by utility allocations generated by *feasible* allocations.

Remark 1 Obviously the definition implies that for $S \neq \emptyset$ the sets $V(S)$ and F are lower comprehensive.

Remark 2 Certain properties of the sets E and $E(S)$ resp. are inherited by the sets F and the $V(S)$ resp.: The sets F and $V(S) \cap \mathbb{R}^S$ are nonempty if E and $\pi_S(E(S))$ are nonempty. Further, F and the $V(S)$ are bounded from above if E and the $\pi_S(E(S))$ are bounded from above.

The further analysis of the properties of F and the $V(S)$ makes it necessary to introduce some additional notation: For $\emptyset \neq S \subset I$ a mapping

$$U^S : \prod_{i \in S} X_i \to \mathbb{R}^S$$

with $U^S(((x_i, P_i))_{i \in S}) := (U_i(x_i, P_I))_{i \in S}$ is defined. The mapping U^S assigns to each S–tuple of consumption plans the corresponding tuple of utility values.

Lemma 4.4 *If the set of feasible allocations E is compact, the set F is closed. If the sets $\pi_S(E(S))$ are compact, the sets $V(S) \cap \mathbb{R}^S$ are closed.*

Proof. The set F can be written as

$$F = U^I(E) + \mathbb{R}^I_-.$$

Being a continuous image of a compact set, the first term is itself a compact set. Since $\mathbb{R}^I_-$ is closed, the asymptotic cones of these sets are positively semi-independent, since the asymptotic cone of a compact set contains only the element zero. Thus the sum of these sets is closed.

Concerning the sets $V(S)$ with $S \neq \emptyset$ the following holds:

$$V(S) \cap \mathbb{R}^S = U^S(\pi_S(E(S))) + \mathbb{R}^S_-.$$

The compactness of the first term follows in the same manner as for the set F. Since R^S_- is obviously closed in $\mathbb{R}^I$, their asymptotic cones are positively semi–independent and, therefore, $V(S) \cap \mathbb{R}^S$ is closed.   □

It can now be shown that given some regularity conditions on the production sets, a production economy with information rules generates a cooperative game. As in chapter two, the set of agents I is identified with the player set. In addition, let F denote the set of feasible payoff allocations and let $V = (V(S))_{S \subset I}$ denote the characteristic function.

Theorem 4.3 *If*

1. *the sets Y and $Y(S)$ are closed in $Map(\Omega, \mathbb{R}^L) \times P^*$ for all $S \subset I, S \neq \emptyset$,*

2. *$(0, Q) \in Y$ and $(0, Q_S) \in Y(S)$ for $Q, Q_S \in P^*$,*

3. *$A(Y)_+ = \{0\} \times P^*$ and $A(Y(S))_+ = \{0\} \times P^*$,*

then

$$(I, F, V)$$

is a cooperative game.

Proof. The theorems 4.1 and 4.2 show that the sets E and $\pi_S(E(S))$ are compact, and lemmas 4.1 and lemma 4.3 imply that they are nonempty. Thus, the sets F and $V(S)$ are nonempty. Remark 1 shows that F and $V(S)$ are lower comprehensive. Lemma 4.4 and Remark 2 imply that the sets F and $V(S) \cap \mathbb{R}^S$ are closed and bounded from above. Thus, all conditions for (I, F, V) being a cooperative game are satisfied. □

Therefore, a coalition production economy generates a cooperative game, provided the sets Y and $Y(S)$ satisfy the requirements stated in theorem 4.3.

4.5 Balanced Technologies and the Existence of Core Allocations

To state conditions which guarantee the non emptiness of the core of a game generated by a production economy, we introduce the concept of a *balanced technology*, as developed by Böhm (1974a, 1974b). As Böhm pointed out, balancedness of a technology is independent of any nonconvexities as caused e.g. by increasing returns to scale or costly disposal.[6] A technology being balanced is a fairly general condition since it implies several other conditions, e.g. the distributivity assumption introduced by Scarf (1986) in a slightly different framework. It is a very general assumption on the technology of an economy, which guarantees the existence of core allocations.

Definition 4.10 *A technology $T = ((Y(S))_{S \subset I}, Y)$ is called **balanced**, if the following condition is satisfied: If $\mathcal{B}$ is a balanced family of coalitions with weights $(\lambda_S)_{S \in \mathcal{B}}$, then*

$$\sum_{S \in \mathcal{B}} \lambda_S Y(S)_Q \subset Y_Q \text{ for all } Q \in P^*.$$

---

[6]'Balancedness neither requires nor imposes properties like free disposal, $0 \in Y(S)$, or non–increasing returns to scale.' Böhm (1988) p.217. As to the properties of balanced technologies see Böhm (1974b).

A technology $T = ((Y(S))_{S \subset I}, Y)$ is thus balanced if the sheets $(Y(S)_Q, Y_Q)$ satisfy the balancedness condition as stated in Böhm (1974a).

The following theorem shows that a cooperative game generated by a production economy is balanced, provided that the technology of the economy is balanced and some additional regularity conditions are satisfied.

Theorem 4.4 *If the technology $T = ((Y(S))_{S \subset I}, Y)$ has the following properties:*

    1. $Y(S)$ and Y are closed in $Map(\Omega, \mathbb{R}^L) \times P^*$ for all $S \subset I$,

    2. $A(Y)_+ = \{0\} \times P^*$,

    3. $(0, Q) \in Y(S)$ for all $S \subset I$ and all $Q \in P^*$,

    4. T is balanced,

and if

    5. $U_i(x_i, p_i)$ is quasi concave for all $i \in I$,

then

$$(I, F, V)$$

is a balanced game.

Proof. First we show that the balancedness of the technology $T = ((Y(S))_{S \subset I}, Y)$ implies the inclusion $Y(S) \subset Y$ for all $S \subset I$.

Given a family of coalitions $\mathcal{B} := \{S, I \setminus S\}$ and weights $\lambda_S := 1 =: \lambda_{I \setminus S}$, $\mathcal{B}$ is a balanced family with weights λ_S. For $S \neq \emptyset$,

$$
\begin{aligned}
Y(S)_Q &= Y(S)_Q + 0 \subset Y(S)_Q + Y(I \setminus S)_Q \\
&= \lambda_S Y(S)_Q + \lambda_{I \setminus S} Y(I \setminus S)_Q \subset Y_Q \text{ for all } Q \in P^*.
\end{aligned}
$$

because of the balancedness of the technology T. Therefore, $Y(S) \subset Y$.

Since $0 \in Y(S)$, this implies that $0 \in Y$. From theorem 4.3, it follows that (I, F, V) is a cooperative game. To prove the balancedness of this game, we have to show that the inclusion

$$\bigcap_{S \in B} V(S) \subset F$$

holds for each balanced family B of subsets of I.

Let $z \in \bigcap_{S \in B} V(S)$ be a payoff allocation. For each $S \in B$ there is an allocation

$$g^S = (x_i^S, P_i^S)_{i \in I} \in E(S)$$

with $U_i(x_i^S, P_i^S) \geq z_i$ for all $i \in S$. Since g^S is an allocation enforceable by the coalition S, it follows that $P_i^S = k_S^i((P_j^0)_{j \in S})$ for all $i \in S$.

Further, there is a production plan $(y^S, Q_S) \in Y(S)$ with $Q_S = t_S((P_i^0)_{i \in S})$, so that:

$$\sum_{i \in S} x_i^S = \sum_{i \in S} e_i + y^S.$$

Using the allocations g^S, a new allocation

$$g = ((x_i, P_i^m))_{i \in I} \in \prod_{i \in I} Abb(\Omega, \mathbb{R}_+^L)$$

and a production plan (y, Q^m) are now constructed as follows:

$$x_i := \sum_{S \in B} \lambda_S x_i^S,$$

$$y := \sum_{S \in B} \lambda_S y^S,$$

where $(\lambda_S)_{S \in \mathcal{B}}$ are the weights of the balanced family $\mathcal{B}$. Since g^S is an allocation enforceable by S, it follows that $x_i^S - e_i \in Map_{P_i^S}(\Omega, \mathbb{R}^L)$, and $y^S \in Map_{Q_S}(\Omega, \mathbb{R}^L)$ for all S. Since the information P_i^m and Q^m are finer than P_i^S and Q_S, it follows that $x_i^S - e_i \in Map_{P_i^m}(\Omega, \mathbb{R}^L)$ and $y^S \in Map_{Q^m}(\Omega, \mathbb{R}^L)$ for all S.

Further,

$$\sum_{i \in I} x_i = \sum_{S \in \mathcal{B}} \lambda_S \sum_{i \in S} x_i^S$$

$$= \sum_{S \in \mathcal{B}} \lambda_S \left[\sum_{i \in S} e_i + y^S \right]$$

$$= \sum_{i \in I} e_i + \sum_{S \in \mathcal{B}} \lambda_S \, y^S.$$

Since the technology T is balanced, it follows that

$$\sum_{S \in \mathcal{B}} \lambda_S \, y(S)_{Q^m} \subset Y_{Q^m}.$$

Thus, there is a production plan $y = \sum_{S \in \mathcal{B}} \lambda_S \, y^S$ in Y_{Q^m}. We have now shown that the equation

$$\sum_{i \in I} x_i = \sum_{i \in I} e_i + y$$

holds, i.e. the allocation is physically feasible. Since $x_i - e_i \in Map_{P_i^m}(\Omega, \mathbb{R}^L)$ and $y \in Map_{Q^m}(\Omega, \mathbb{R}^L)$, i.e. the net trades as well as the production plans are also compatible with the maximum amount of information, the allocation $g = ((x_i, P_i)_{i \in I}$ is feasible. Thus, the utility allocation $(U_i(x_i, P_i))_{i \in I}$ associated with g is an element of F.

If it is shown that this utility allocation dominates the payoff allocation

$$z \in \bigcap_{S \in \mathcal{B}} V(S),$$

it follows that $z \in F$.

Since preferences are convex and the x_i are convex combinations of the x_i^S, we have

$$U_i(x_i, P_i) \geq z_i \quad \forall i \in I.$$

Thus, the game (I, F, V) is balanced. $\qquad \square$

The existence of core allocations immediately follows as a simple corollary.

Corollary 4.1 *Let $\mathcal{E}$ be an economy*

$$\mathcal{E} = \left(I, \left(X_i, U_i, e_i, P_i^0\right)_{i=1}^{|I|}, T = ((Y(S))_{S \subset I}, Y), k, t \right).$$

If the conditions 1 - 5 of theorem 4.4 are satisfied, then the core of the economy $\mathcal{E}$ is nonempty.

Proof. Theorem 4.4 implies that the associated game (I, F, V) is balanced and has a nonempty core as is shown in the theorem of Böhm (1974a). Since each payoff allocation $z \in F$ is dominated by a utility allocation of a feasible allocation in $\mathcal{E}$, there is an allocation $g = ((x_i, P_i))_{i \in I}$ in $\mathcal{E}$ with $(U_i(x_i, P_i))_{i \in I} \geq z$. Thus, g is an allocation in the core of the economy $\mathcal{E}$. $\qquad \square$

In what follows, we will illustrate the impact of asymmetric information on the core of a production economy by a simple example.

Example 4.1 Let there be two states of nature, i.e. $\Omega = \{\omega_1, \omega_2\}$, and two agents. The initial information of the agents is given by $P_i^0 := \{\{\omega_1\}, \{\omega_2\}\}$, $i = 1, 2$. There are two contingent commodities, i.e. the commodity space is given by $\mathbb{R}^2$. The consumption sets are $X_i = \mathbb{R}_+^2$. The initial endowments are given by $e_1 := (9, 3)$

and $e_2 := (63, 6)$ and the agents' preferences are described by the utility functions $U_i(x_i) := x_{i1}x_{i2}$, $i = 1, 2$.

Each coalition is endowed with a technology to produce commodity 2 by using commodity 1 as an input. The technology sets $Y^{\{1\}}, Y^{\{2\}}, Y^{\{1,2\}}$ are identical and described by the production functions $f_{\{1\}}(y_1) = f_{\{2\}}(y_1) = f_{\{1,2\}}(y_1) := y_1^{1/2}$. The set Y is given by $Y = \sum_{S \subset I} Y^S$. Obviously, the technology of the economy $(Y(S)_{S \subset I}, Y)$ is balanced.[7]

With respect to the information rules, it is assumed that $k_S := id$ and $t_S := \Omega^*$ holds for all $S \subset I$. Agents have access to complete information for their transactions as well as for production.

To be an element of the core of this economy with the information rules k and t, an allocation (x_1, x_2) has to satisfy the following three conditions:

1. $x_1 + x_2 = e_1 + e_2 + y$, with $y \in Y$ (physical feasibility),

2. $MRS_{\{1\}}(x_1) = MRS_{\{2\}}(x_2) = MRT_{\{1,2\}}(e_{11} + e_{21} - y_1, e_{12} + e_{22} + y_1^{1/2})$ (Pareto-optimality),

3. $U_i(x_i) \geq U_i(\bar{x}_i)$, where $\bar{x}_i = e_i + y_i$ with $y_i \in Y^{\{i\}}$ and $MRS_{\{i\}}(\bar{x}_i) = MRT_{\{i\}}(e_{i1} - y_{i1}, e_{i2} + y_{i1}^{1/2})$ (individual rationality).

Here, MRS_S and MRT_S denote the marginal rates of substitution and transformation for $S \subset I$.

Notice that there are no constraints on informational feasibility, since complete information ensures that there are no measurability constraints to be taken into account.

Condition 2 implies

$$y_1 = 3$$

---

[7]See Böhm (1974b) p.12.

and condition 3 implies $y_{11} = 1$ and $y_{21} = 3$.

Individually rational allocations have to satisfy the conditions

$$U_1(x_{11}, x_{12}) \geq 32,$$

and for agent 2:

$$U_2(x_{21}, x_{22}) \geq 360 + 60 \cdot \sqrt{3} \approx 464.$$

The Pareto-optimal allocations (x_1, x_2) satisfy

$$x_{12} = \frac{9 + \sqrt{3}}{69} x_{11}.$$

Thus, for allocations in the core of the economy it must be that $14,344 \leq x_{11} \leq 14,381$. Core allocations are characterized by three units of commodity 1 being used as input in production.

To illustrate the impact of asymmetric information, it is now assumed that the information the grand coalition can use in production is determined by $t_I := \{\Omega\}$, i.e. it can choose only state independent production plans. But since production in this economy is equivalent to shifting the physical commodity from one state of nature to the other, there is only one production plan y that satisfies this condition, i.e. $y = 0$. Stated otherwise, the grand coalition is unable to produce and can only reallocate the endowments of its members.

However, the individually rational allocations cannot be blocked by the grand coalition using a mere reallocation of its members' endowments. If this were possible, the inequalities

$$
\begin{aligned}
x_{11}x_{12} &\geq 32, \\
(72 - x_{11})(9 - x_{12}) &\geq 464
\end{aligned}
$$

would have to be satisfied simultaneously. But the set of solutions is empty, as can be seen in the following diagram.

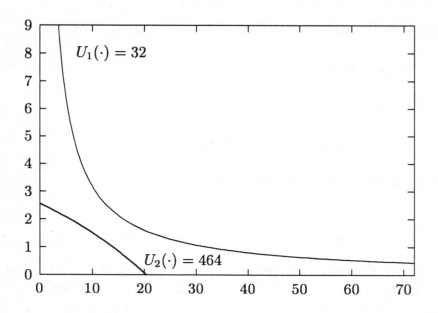

Consider an allocation where 4 units of commodity 1 are trans-
formed into commodity 2. Such an allocation is an element of the
core only if the information is asymmetric. Or, stated otherwise:
Allocations that exhibit 'overproduction' (as compared to the situation of complete information) are unblocked by any coalition.
This example demonstrates that asymmetric information can indeed be an incentive for decentralization.

4.6   Costs of Coalition Formation

Up until now it has been assumed that each coalition is able
to do without production and can only reallocate its members
endowments. This was independent of the information that can be
used in production. Formally, it was assumed that $(0, Q) \in Y(S)$
for all $S \subset I$ and all $Q \in P^*$. Under this assumption, the set of
allocations enforceable by a coalition is always nonempty since an

allocation that gives each member his initial endowment is always enforceable.

However, as mentioned above, one can imagine that coalition formation is costly. This is always the case if resources have to used to make it possible that agents can form a coalition as pointed out by Böhm (1974a): 'This includes the case where there are costs of forming the coalition as well as reaching a joint decision.'[8]

In this case, the production plan $(0, Q)$ will in general not be a feasible production plan for the coalition, i.e. is not in the coalitions' technology set. It even might happen that the real costs of coalition formation are so high that they are larger than the total initial endowments of the coalition members. In this case there are no allocations whatsoever which this coalition can enforce, i.e. $E(S) = \emptyset$.

In what follows, the analysis is extended to the case where the sets of enforceable allocations might be empty. However, it is assumed that singleton coalitions, where there are no costs of coalition formation, are always able to consume their initial endowments. Formally, it is assumed that $E(\{i\}) \neq \emptyset$ for all $i \in I$. It will be shown that the more general case where the sets $E(S)$ might be empty can be reduced to the case where all $E(S)$ are nonempty.

Recall that a subset $M \subset Map(\Omega, \mathbb{R}^L) \times P^*$ can be written as

$$
\begin{aligned}
M &= \bigcup_{Q \in P^*} \left(M \cap (Map(\Omega, \mathbb{R}^L) \times \{Q\}) \right) \\
&= \bigcup_{Q \in P^*} \left(M_Q \times \{Q\} \right),
\end{aligned}
$$

i.e. M is given by the family

$$
(M_Q)_{Q \in P^*} \subset \prod_{Q \in P^*} Map(\Omega, \mathbb{R}^L).
$$

In words: the set M can be reconstructed from the sheets M_Q.

---

[8] Böhm (1974a) p.430

Definition 4.11 *Let $T = ((Y(S))_{S \subset I}, Y)$ be a technology and let $\tilde{S} \subset I$ be a given coalition. Using this technology, a new technology $T_{\tilde{S}} = \left((\tilde{Y}(S))_{S \subset I}, Y \right)$ is constructed. This is done by replacing $Y(S)_Q$ by $\tilde{Y}(S)_Q$ which is given by*

$$\tilde{Y}(S)_Q := \begin{cases} Y(S)_Q & S \neq \tilde{S}, \\ \sum_{i \in \tilde{S}} Y(\{i\})_Q & S = \tilde{S}, \end{cases}$$

for all $Q \in P^*$.

Stated otherwise, the technology $T_{\tilde{S}}$ differs from the technology T by replacing $Y(\tilde{S})_Q$ by $\tilde{Y}(\tilde{S})_Q = \sum_{i \in \tilde{S}} Y(\{i\})_Q$. Notice that for the description of a technology it is sufficient to consider each sheet of the technology separately.

The following lemma shows that the property of balancedness of the technology T is inherited by the technology $T_{\tilde{S}}$.

Lemma 4.5 *The technology $T_{\tilde{S}}$ is balanced if the technology T is balanced.*

Proof. Let $\mathcal{B}$ be a balanced family of subsets of I with balancing weights $(\lambda_S)_{S \in \mathcal{B}}$. It has to be shown that for all $Q \in P^*$, it is true that

$$\sum_{S \in \mathcal{B}} \lambda_S \tilde{Y}(S)_Q \subset Y_Q.$$

If $\tilde{S} \notin \mathcal{B}$, then

$$\sum_{S \in \mathcal{B}} \lambda_S \tilde{Y}(S)_Q = \sum_{S \in \mathcal{B}} \lambda_S Y(S)_Q \subset Y_Q,$$

since T is balanced.

Therefore let $\tilde{S} \in \mathcal{B}$. The family of subsets of I, $\tilde{\mathcal{B}}$, is defined by

$$\tilde{\mathcal{B}} := \mathcal{B} \setminus \{\tilde{S}\} \cup \left\{ \{i\} \mid i \in \tilde{S} \right\},$$

and the balancing weights $\left(\tilde{\lambda}_S \right)_{S \in \tilde{\mathcal{B}}}$ are given by

$$\tilde{\lambda}_S := \begin{cases} \lambda_S & S \in \mathcal{B} \setminus \{\tilde{S}\}, \\ \lambda_{\tilde{S}} & S \in \left\{ \{i\} \mid i \in \tilde{S} \right\}. \end{cases}$$

Thus, $\tilde{\lambda}_{\{i\}} = \lambda_{\tilde{S}}$ for all $i \in \tilde{S}$. $\tilde{\mathcal{B}}$ is a balanced family with weights $\left(\tilde{\lambda}_S \right)_{S \in \tilde{\mathcal{B}}}$ since:

$$\begin{aligned} \sum_{S \in \tilde{\mathcal{B}}} \tilde{\lambda}_S \chi_S &= \sum_{S \in \mathcal{B} \setminus \{\tilde{S}\}} \tilde{\lambda}_S \chi_S + \sum_{i \in \tilde{S}} \tilde{\lambda}_{\{i\}} \chi_{\{i\}} \\ &= \sum_{S \in \mathcal{B} \setminus \{\tilde{S}\}} \lambda_S \chi_S + \lambda_{\tilde{S}} \sum_{i \in \tilde{S}} \chi_{\{i\}} \\ &= \sum_{S \in \mathcal{B} \setminus \{\tilde{S}\}} \lambda_S \chi_S + \lambda_{\tilde{S}} \chi_{\tilde{S}} \\ &= \sum_{S \in \mathcal{B}} \lambda_S \chi_S = \chi_I, \end{aligned}$$

because $\mathcal{B}$ is balanced with weights $(\lambda_S)_{S \in \mathcal{B}}$. Here χ_S denotes the characteristic function of the subset $S \subset I$, i.e.

$$\chi_S(x) := \begin{cases} 1 & x \in S, \\ 0 & x \notin S. \end{cases}$$

It follows that

$$\sum_{S \in \tilde{\mathcal{B}}} \tilde{\lambda}_S Y(S)_Q \subset Y_Q$$

for all $Q \in P^*$, since the technology T is balanced. For the technology $T_{\tilde{S}}$ it follows that:

$$
\begin{aligned}
\sum_{S \in B} \lambda_S \tilde{Y}(S)_Q &= \sum_{S \in B \setminus \{\tilde{S}\}} \lambda_S \tilde{Y}(S)_Q + \lambda_{\tilde{S}} \tilde{Y}(\tilde{S})_Q \\
&= \sum_{S \in B \setminus \{\tilde{S}\}} \tilde{\lambda}_S Y(S)_Q + \lambda_{\tilde{S}} \sum_{i \in S} Y(\{i\})_Q \\
&= \sum_{S \in B \setminus \{\tilde{S}\}} \tilde{\lambda}_S Y(S)_Q + \sum_{i \in \tilde{S}} \tilde{\lambda}_{\{i\}} Y(\{i\})_Q \\
&= \sum_{S \in \tilde{B}} \tilde{\lambda}_S Y(S)_Q \subset Y_Q
\end{aligned}
$$

for all $Q \in P^*$. Thus, the technology $T_{\tilde{S}}$ is balanced.   □

In what follows, these concepts will be used to analyze the core of an economy where coalitions can enforce no allocations whatsoever, i.e. where the sets of enforceable allocations are empty. As a first step for a given economy a technology $\tilde{T} = \left((\tilde{Y}(S))_{S \subset I}, Y \right)$ is constructed from the balanced technology $T = ((Y(S))_{S \subset I}, Y)$. The new technology differs from the former only with respect to those coalitions which cannot enforce any allocations in the economy.

Theorem 4.5 *If the technology $T = ((Y(S))_{S \subset I}, Y)$ is balanced with $0 \in Y(\{i\})_Q$ for all $Q \in P^*$ and for all $i \in I$, than there is a balanced technology $T_{\tilde{S}} = \left((\tilde{Y}(S))_{S \subset I}, Y \right)$, such that $E_{\tilde{T}}(S) \neq \emptyset$ for all $S \subset I$ and $\tilde{Y}(S) = Y(S)$ if $E_T(S) \neq \emptyset$. Here $E_{\tilde{T}}(S)$ denotes the set of allocations enforceable by $S \subset I$ for the technology $\tilde{T}$.*

Proof. Let $\mathcal{D}$ be the set of all coalitions without enforceable allocations, i.e. $\mathcal{D} = \{S \subset I \mid E_T(S) = \emptyset\}$, and let $S_1 \in \mathcal{D}$. Let the technology $T_{S_1} = ((Y_1(S))_{S \subset I}, Y)$ be given as in definition 4.11. Since $Y_1(S)_Q = \sum_{i \in S_1} Y(\{i\})_Q$ and $0 \in Y(\{i\})_Q$ for all $Q \in P^*$

and all $i \in S_1$, it is obvious that $((e_i, P_i^0))_{i \in I} \in E_{T_{S_1}}(S_1)$. Thus it follows that $E_{T_{S_1}}(S_1) \neq \emptyset$. Further, if

$$\mathcal{D}_1 := \{S \subset I \mid E_{T_{S_1}}(S) = \emptyset\},$$

then

$$\mathcal{D}_1 = \mathcal{D} \setminus \{S_1\}.$$

Now the existence of $\tilde{T} = \left((\tilde{Y}(S))_{S \subset I}, Y\right)$ follows by induction.
$\square$

The following lemma states that the sets of enforceable allocations $E_{\tilde{T}}(S)$, given the technology $\tilde{T}$, differ at most with respect to the coalitions which do not have any enforceable allocations given the technology T.

Lemma 4.6 *Let $T = ((Y(S))_{S \subset I}, Y)$ be a balanced technology with $0 \in Y(\{i\})_Q$ for all $Q \in P^*$ and all $i \in I$. Further, let $\tilde{T} = \left((\tilde{Y}(S))_{S \subset I}, Y\right)$ denote the technology as given in theorem 4.5 with $E_{\tilde{T}}(S) \neq \emptyset$ for all $S \subset I$. Than it holds that*

$$E_T(S) = E_{\tilde{T}}(S) \text{ if } E_T(S) \neq \emptyset.$$

Proof. Let $((x_i, P_i))_{i \in I} \in E_T(S) \neq \emptyset$, i.e.

$$\sum_{i \in S} x_i = \sum_{i \in S} e_i + y \text{ with } (x_i, P_i) \in X_i \text{ and } (y, Q) \in Y(S),$$

where $P_i = k_S^i\left((P_j^0)_{j \in S}\right)$ and $Q = t_S\left((P_j^0)_{j \in S}\right)$. Since $\tilde{Y}(S) = Y(S)$ for $E_T(S) \neq \emptyset$ it holds that $((x_i, P_i))_{i \in I} \in E_{\tilde{T}}(S)$.
$\square$

Using theorem 4.5 and lemma 4.6 it can now be shown that the core of a coalition production economy is nonempty even if – due to high costs of coalition formation – some coalitions don't have any enforceable allocations whatsoever.

Theorem 4.6 *Let $\mathcal{E}$ be an economy*

$$\mathcal{E} = \left(I, (X_i, U_i, e_i, P_i^0)_{i=1}^{|I|}, T = ((Y(S))_{S \subset I}, Y), k, t \right).$$

If

1. *$Y(S)$ and Y are closed in $Map(\Omega, \mathbb{R}^L) \times P^*$ for all $S \subset I$,*

2. *$A(Y)_+ = \{0\} \times P^*$,*

3. *$0 \in Y(\{i\})_Q$ for all $i \in I$ and all $Q \in P^*$,*

4. *T is balanced,*

5. *$U_i(x_i, P_i)$ is quasi–concave for all $i \in I$,*

and if $\tilde{\mathcal{E}}$ is the economy derived from $\mathcal{E}$ by replacing the technology T by $\tilde{T}$ as given in theorem 4.5, than it follows that

$$Core(\tilde{\mathcal{E}}) \subset Core(\mathcal{E}).$$

In particular, $Core(\mathcal{E}) \neq \emptyset$.

Proof. Let $g = ((x_i, P_i))_{i \in I} \notin Core(\mathcal{E})$, i.e. there is a coalition $S \subset I$ and an allocation $g' = ((x_i', P_i'))_{i \in I} \in E_T(S)$, such that $U_i(x_i', P_i') > U_i(x_i, P_i)$ for all $i \in S$. In particular, $E_T(S) \neq \emptyset$, such that $E_T(S) = E_{\tilde{T}}(S)$ follows from lemma 4.6. This implies $g \notin Core(\tilde{\mathcal{E}})$. $\square$

But even in the case that some coalitions cannot enforce any allocations, only those allocations are in the core where every agents gets at least a utility that is associated with his initial endowment. Otherwise, an agent who receives a lower utility could block such an allocation by forming a singleton coalition and consuming his initial endowment or produce in autarky. Core allocations have the property of individual rationality even if coalitions have no enforceable allocations.

4.7 Summary and Discussion

In this chapter the core of a coalition production economy with asymmetric information is analyzed. The production possibilities for each coalition are described by assigning to every coalition a technology set and the information the coalition can employ in the production process. The information a coalition can use in production is modelled – as in chapter 3 – by an information rule. In contrast to the information the members of a coalition can use to carry out reallocations, the production information rule assigns only *one* information to each coalition, i.e. one partition of the set of states of nature. The reason is that a coalition's production decision is considered a *joint* decision, based on one information. The technology sets are assumed to be closed subsets of the commodity space. In addition, only those production plans are considered as feasible which are compatible with the information which can be used in production.

In order for the core to be nonempty, the asymptotic cone of a technology set has to satisfy a certain property. Since the commodity space is not a Euclidean space, the concept of an asymptotic cone has to be extended according to the new commodity space. It is not assumed that the technology sets are convex, i.e. increasing returns to scale are allowed for. However, it is assumed that the technology of the economy, i.e. the set of technology sets, is balanced. This concept has been introduced by Böhm (1974a) and is the most general assumption which guarantees the existence of core allocations in a coalition production economy.

Under these assumptions – modified with respect to informational considerations – it can be shown that even if agents or coalitions are asymmetrically informed, the core of such an economy is nonempty. This holds true even if forming a coalition is costly so that for some coalitions there are no enforceable allocations whatsoever. The existence of core allocations is guaranteed independently of the information rules which determine the information that can be used for reallocations and for production decisions.

The impact of asymmetric information is illustrated by a simple example. This example shows that asymmetric information may have important consequences for the production decisions in the core of an economy. In the case of asymmetric information the core contains allocations which would be blocked by the grand coalition if information were perfect. In this example, the information of the grand coalition is chosen in such a way that only no production at all is compatible with this information. This implies that the grand coalition can enforce only those allocations which can be achieved by reallocations of the coalition members' initial endowments. These allocations, however, are blocked by the singleton coalitions, since an individual production guarantees a higher utility to the agents, as compared to the utility which can be achieved by a reallocation of the endowments. But by producing individually, more resources are used as in the case of perfect information. This shows that asymmetric information may lead to core allocations which – compared to core allocations with perfect information – imply an 'overproduction'. Asymmetric information may also be an incentive for decentralization of production, a conjecture which has been made already by Böhm (1974a).

A priori, we do not impose any relationship between the information rules concerning the information to be used for transactions on the one hand and for the information used in production on the other hand. The reason for treating these two types of information separately is of a conceptual nature since each agent plays a dual role: he is a consumer and a producer at the same time. There is a priori no direct relationship between these two functions. However, one could imagine that, if the members of a coalition were able to pool their information to make transactions, the same aggregation should also be possible concerning the production decisions. In this case, the information which can be used in production would be directly related to the information which the agents can employ for transactions. Since the analysis applies to all combinations of information rules, the main results of the paper still holds true in these cases.

Another problem which has not been discussed is the relationship between the core of an economy with asymmetric information and a Walras equilibrium in this economy. For example, one could analyze the question of whether a Walras equilibrium is always an element of the core, or if it could happen that in the case of asymmetric information a Walras equilibrium is not in the core. Introducing prices into the model, one could also consider *stable firm structures*, a conceptual framework which has independently been developed by Böhm (1973) and Sondermann (1974). A stable firm structure allows for the endogenous determination of the number of firms which are active on markets. This approach could be interpreted as a model of market entry and exit and allows for the incorporation of some approaches from the theory of the firm as well as from the theory of industrial organization in a model of general economic equilibrium.

Conclusion

The focal point of the work presented here was the development of a conceptual framework which allows the consistent integration of information in a model of general equilibrium. Further, it was the analysis whether and under which conditions the core of economies with asymmetric information is nonempty. Using the concept of the core as equilibrium concept, some ideas from contract theory, a partial equilibrium framework – i.e. the conclusion of binding agreements between economic subjects – can be transferred to a model of an economy. However, the models presented here are just a first step to analyze the impact of informational problems on an economy and can be extended in several respects. Some of these extensions are sketched in the following. First, it is the integration of strategic information transmission, the modeling of Walras equilibria with asymmetric information and its relationship to the core, and the problem of stable firm structures, an concept developed independently by Böhm (1973) and Sondermann (1974).

The modeling of information exchange used here differs from the approach employed in contract theory. In the latter, transmission of information is determined as an equilibrium in a noncooperative game. The exogenously given information rules used here abstract from any strategic information transmission. A possible extension of the model presented here would be an explicit consideration of a strategic information transmission between the members of a coalition. Up until now this aspect has not been considered in the literature. The only paper which analyzes this problem has been presented by Ichiishi, Idzak and Zhao[9], which however employ a different theoretical framework. They introduce the concept of a *social coalitional equilibrium*, a synthesis of the Nash–equilibrium of a generalized game and the core. The coalition members' decisions about information transmission are determined endogenously in this model. It would be worthwhile to examine whether how far results about strategic information transmission in coalitions can be transferred to the model presented here.

Another aspect which has not been discussed here is the relationship between the core of an economy with asymmetric information and the Walras equilibrium of such an economy. Here the following problem arises: The commodity space considered here is a non Euclidean space. Thus, it is not clear how prices and therefore Walras equilibria have to be defined. For a usual commodity space, prices are defined as a linear functional on this space, which cannot be done in the case considered here. Nevertheless, for each sheet of the commodity space linear functionals can be defined. This allows to define prices separately for each sheet of the commodity space. A Walras equilibrium of an exchange economy could then be defined as a tuple of price vectors – one for each sheet – and an allocation with the properties that, given the initial endowments, each consumer maximizes his utility and all markets are cleared. The central questions in this context refer above all to the existence and the properties of such a Walras equilibrium. Assuming the existence of an equilibrium, a reasonable conjecture is that economic agents who differ only with respect to their information will in equilibrium pay different prices for the state contingent commodity – a phenomenon which is often observed in reality. Of course this leads directly to questions referring to the relationship between the core and the Walras equilibrium of an economy with asymmetric information. For example, it could be examined which information rules lead to a Walras equilibrium being an element of the core of the economy.

In a further step the problem of asymmetric information in large

---

[9]See Ichiishi, Idzak, Zhao (1994).

economies could be analyzed. Here, the question arises whether the core of the economy shrinks if the economy is replicated. If this is the case, it would be interesting to find out which allocations are in the core of the limit economy. A reasonable conjecture would be that in large economies the impact of asymmetric information depends on the opportunities of information exchange. As discussed in section 2.12, Srivastava has shown that for information rules where the information of a coalition member increases in coalition size, the core of the economy will shrink and will converge to the core of an economy with perfect information. However, no statement is made with respect to core allocations in replica economies if other information rules are given.

The Introduction of prices into the conceptual framework of the model as outlined above would allow for the analysis of stable firm structures in an economy with asymmetric information. As mentioned above, this concept has been introduced independently by Böhm (1973) and Sondermann (1974) and allows for the endogenous determination of the number of firms which are active in markets. This can be interpreted as a model of market entry and exit. Such a model allows for the integration of models of the firm and of industrial organization, which usually assume asymmetric information, into a general equilibrium framework.

List of Symbols

I	the set of agents in the economy
i	an agent
Ω	the set of states of nature
$\mathcal{P}(\Omega)$	the powerset of Ω
P^*	the set of all partitions of Ω
Ω^*	the singleton partition of Ω
P_i	an information of agent i
P_i^0	the initial information of agent i
$(P_i)_{i \in I}$	an information structure of the economy
$P_i(\omega)$	the element of P_i containing ω
$\sigma(P_i)$	the sigma algebra generated by P_i
$\mathbb{R}^L$	the L–dimensional Euclidean space
$Map(\Omega, \mathbb{R}^L)$	the contingent commodity space
$Map(\Omega, \mathbb{R}^L) \times P^*$	the generalized commodity space
e_i	the initial endowment of contingent commodities of agent i
(e_i, P_i^0)	the initial endowment of agent i
X_i	the consumption set of agent i
(x_i, P_i)	a consumption plan of agent i
$U_i(x_i, P_i)$	the utility function of agent i
$(x_i, P_i)_{i \in I}$	an allocation
$\mu_i(\omega)$	the subjective probability of event ω of agent i
$S \subset I$	the coalition S
k_s	an information rule for coalition S
Id	the identity mapping
$\mathcal{P}(I)$	the set of all coalitions

$(k_s)_{S \in \mathcal{P}(\mathcal{I})}$	an information rule for coalition S
k_S^c	the coarse information rule
k_S^f	the fine information rule
k_S^p	the private information rule
k_S^n	the null information rule
$\mathcal{K}$	the set of all information rules
$\mathcal{E}$	an economy
$\mathcal{E}^{\|}$	an economy with information rule k
P_i^m	the maximum amount of information of agent i
E	the set of feasible allocations
F	the set of feasible payoff allocations
F^k	the set of feasible payoff allocations given information rule k
V	the characteristic function
z	a payoff vector
$A(M)$	the asymptotic cone of set M
$\mathcal{B}$	a familiy of subsets of I
λ_S	the balancing weight for coalition S
t^s	a production information rule
Q_S	the information coalition S can employ in production
$Y(S)$	the technology set of coalition S
Y	total technology set of the economy
y	a state contingent production plan
T	the technology of the economy
$\mathcal{E}^{kt}$	a coalition production economy with rules k and t
M_+	the nonnegative elements of a set M
π	the projection on the first component
$E_{(P_i)_{i \in I}}$	the set of feasible state contingent commodities
U^S	a utility allocation for coalition S
χ_S	the indicator function of coalition S
$\mathcal{D}$	the set of coalitions without enforceable allocations

References

AKERLOF, G.A. (1970), *The Market for 'Lemons': Qualitative Uncertainty and the Market Mechanism*, Quaterly Journal of Economics, 84, p.488-500

ALLEN B. (1981), *Generic Existence of Completely Revealing Equilibria for Economies when Prices Convey Information*, Econometrica, 49, p.1173-1199

ALLEN B. (1986), *General Equilibrium with Information Sales*, Theory and Decision, 21, p.1-33

ALLEN B. (1990), *Information as an Economic Commodity*, American Economic Review, Papers and Proceedings, 80, p.68-273

ALLEN B. (1991a), *Market Games with Asymmetric Information and Nontransferable Utility: Representation Results and the Core*, University of Pennsylvania, CARESS Working Paper 91-09

ALLEN B. (1991b), *Transferable Utility Market Games with Asymmetric Information: Representation Results and the Core*, University of Pennsylvania, CARESS Working Paper 91-16

ALLEN B. (1991c), *Incentives in Market Games with Asymmetric Information: The Core*, University of Pennsylvania, CARESS Working Paper 91-38

ALLEN, B. (1992), *Market Games with Asymmetric Information: The Private Information Core*, University of Pennsylvania, CARESS Working Paper 92-04

ALLEN, B. (1994a), *Market Games with Asymmetric Information: The Core with Finitely Many States of the World*, in *Models and Experiments in Risk and Rationality*, eds. B. Munier and M.J. Machina, Kluwer Academic Press

ALLEN, B. (1994b), *Incentives in Market Games with Asymmetric Information: Approximate NTU Cores in Large Economies*, in *Social Choice, Welfare and Ethics*, eds. W. Barnett, H. Moulin, M. Salles and N. Schofield, Cambridge University Press

ALLEN, B. (1995), *On the Existence of Core Allocations in a Large Economy with Incentive–Compatibility Constraints*, University of Minnesota and Federal Reserve Bank of Minneapolis, mimeo

ARROW, K. (1973), *Information and Economic Behaviour*, in: Arrow, K.: Essays in the Theory of Risk Bearing, Amsterdam, Oxford; North-Holland, p.136-152

BERLIANT, M. (1992), *On Income Taxation and the Core*, Journal of Economic Theory, 56, p.121-141

BÖHM, V. (1973), *Firms and Market Equilibria in a Private Ownership Economy*, Zeitschrift für Nationalökonomie, 33, p.87-102

BÖHM, V. (1974a), *The Core of an Economy with Production*, Review of Economic Studies, XLV, p.429-436

BÖHM, V. (1974b), *On Balanced Games, Cores and Production*, Core Discussion Paper No. 7430, Université Catholique de Louvain

BÖHM, V. (1988), *Returns to Size vs. Returns to Scale: The Core with Production Revisited*, Journal of Economic Theory, 46, p.215-219

BOYD, J., E. PRESCOTT (1986), *Financial Intermediary–Coalitions*, Journal of Economic Theory, 38, p.211-232

COTTER, K.D. (1986), *Similarity of Information and Behavior with a Pointwise Convergence Topologie*, Journal of Mathematical Economics, 15, 25-38

DASGUPTA, P., J. STIGLITZ (1980), *Uncertainty, Industrial Structure and the Speed of R & D*, Bell Journal of Economics, 11, p.1-28

DEBREU, G. (1959), *Theory of Value*, Cowles Foundation Monograph 17, New York, Wiley

ELLICKSON, B. (1993), *Competitive Equlibrium*, Cambridge; Cambridge University Press

GALE, D. (1978), *The Core of a Monetary Economy without Trust*, Journal of Economic Theory, 19, 456-491

GUL, F., A. POSTLEWAITE (1992), *Asymptotic Efficiency in Large Exchange Economies with Asymmetric Information*, Econometrica, 60, p.1273-1292

GUESNERIE, R., TH. DE MONTBRIAL (1974), *Allocation under Uncertainty: A Survey*, in: Allocation under Uncertainty: Equilibrium and Optimality (ed. J.H. Dréze), London; Macmillan

HAHN, G., N.C. YANNELIS (1995), *Coalitional Bayesian Nash Implementation in Differential Information Economies*, Department of Economics, University of Illinois, Urbana–Champaign, mimeo

HART, O., B. HOLMSTRÖM (1987) *The Theory of Contracts*, in: Advances in Economic Theory, Fifth World Congress, (ed. T.F. Bewley), Cambridge; Cambridge University Press, p.71-155

HAYEK, F. VON , (1937) *Economics and Knowledge*, Economica IV, new series, p.33-54

HILDENBRAND, W. (1974) *Core and Equilibria of a Large Economy*, Princeton; Princeton University Press

HILDENBRAND, W., A. KIRMAN (1988), *Equilibrium Analysis*, North-Holland, Amsterdam, New York, Tokyo

HIRSHLIFER, J., J.G. RILEY (1992), *The Analytics of Uncertainty and Information*, Cambridge; Cambridge University Press

HONKAPOHJA, S. (1977), *Money and the Core in a Sequence Economy with Transaction Costs*, European Economic Review, 10, 241-251

ICHIISHI, T. (1993), *The Cooperative Nature of the Firm*, Cambridge; Cambridge University Press

ICHIISHI, T., A. IDZAG, J. ZHAO (1994), *Cooperative Processing of Information Via Choice at an Information Set*, International Journal of Game Theory, 23, p.145-166

JORDAN, J.S., R. RADNER (1982), *Rational Expectations in Microeconomic Models: An Overview*, Journal of Economic Theory, 26, p.201-223

KOUTSOUGERAS, L. (1998), *A Two Stage Core with Applications to Asset Market and Differential Information Economies*, Economic Theory, 11, p.563-584

KOUTSOUGERAS, L., N.C. YANNELIS (1994), *Incentive Compatibility and Information Superiority of the Core of an Economy with Differential Information*, Economic Theory, 3, p.195-2169

KOUTSOUGERAS, L., N.C. YANNELIS (1995), *Learning in Differential Information Economies with Cooperative Solution Concepts: Core and Value*, Core Discussion Paper 9545, Louvain–La–Neuve

KRASA, S., N.C. YANNELIS (1994), *The Value Allocation of an Economy with Differential Information* Econometrica, 62, p.881-900

LEE, D (1998), *Essays on the Core of Economies with Asymmetric Information*, Ph.D. Thesis, Brown University, Providence, RI

MARIMON, R. (1989), *The Core of Private Information Economies*, UAB/IAE Discussion Paper No. 131.90, Universitat Autonoma de Barcelona

MILGROM, P., J. ROBERTS (1992), *Economics, Organization and Management*, Englewood Cliffs; Prentice Hall

MOORE, J. (1990), *Implementation, Contracts, and Renegotiation in Environments with Complete Information*, in: Advances in Economic Theory, Sixth World Congress Vol.I, (ed. J.-J. Laffont), Cambridge; Cambridge University Press, p.183-282

PALFREY, TH.R. (1990), *Implementation in Bayesian Equilibrium: The Multiple Equilibrium Problem in Mechanism Design*, in: Advances in Economic Theory, Sixth World Congress Vol.I, (ed. J.-J. Laffont), Cambridge; Cambridge University Press, p.283-327

PELEG, B. (1985), *The Axiomatization of the Core of Cooperative Games without Side Payments*, Journal of Mathematical Economics, 14, 203-214

PHLIPS, L. (1988), *The Economics of Imperfect Information*, Cambridge; Cambridge University Press

RADNER, R. (1968), *General Equilibrium under Uncertainty*, Econometrica, 36, p.31-58

RADNER, R. (1972), *Existence of Equilibrium of Plans, Prices and Price Expectations in a Sequence of Markets*, Econometrica, 40, p.289-304

RADNER, R. (1982), *Equilibrium under Uncertainty*, in: Handbook of Mathematical Economics Vol. III, (eds. Arrow, K.J., M.D. Intrilligator), Amsterdam, New York, Oxford; North-Holland, p.923-1006

REINGANUM, J. (1983), *Technology Adoption under Imperfect Information*, Bell Journal of Economics, 14, p.57-69

REPULLO, R. (1988), *The Core of an Economy with Transaction Costs*, Review of Economic Studies, 55, 447-458

SALANIÉ, B. (1997), *The Economics of Contracts*, Cambridge, Mass., MIT Press

SCARF, H. (1967), *The Core of an n-person Game*, Econometrica, 35, p.50-69

SONDERMANN, D. (1974), *Economies of Scale and Equilibria in Coalition Production Economies*, Journal of Economic Theory, 8, p.259-291

SRIVASTAVA, S. (1984a), *Rational Expectations Equilibria and the Core*, Carnegie-Mellon University, Working Paper 1382-83

SRIVASTAVA, S. (1984b), *A Limit Theorem on the Core with Differential Information*, Economics Letters, 14, p.111-116

STIGLER, G.J. (1961), *The Economics of Information*, Journal of Political Economy, 69, p.213-225

STIGLITZ, J. (1984), *Information and Economic Analysis: A Perspective*, Economic Journal, 95, Conference Papers, p.21-41

STINCHCOMBE, M.B. (1990), *Bayesian Information Topologies*, Journal of Mathematical Economics, 19, 233-253

TAYLOR, M.P. (1987), *The Simple Analytics of Implicit Labour Contracts*, in: Surveys in the Theory of Uncertainty, (eds Hey, J.D., P.J. Lambert), Oxford; Blackwell, p.124-150

VOHRA, R. (1997), *Incomplete Information, Incentive Compatibility and the Core*, Economics Department, Brown University, Providence, RI, mimeo

VOLIJ, O. (1997), *Communication, Credible Improvements and the Core of an Economy with Asymmetric Information*, Economics Department, Brown University, Providence, RI, mimeo

WILSON, R. (1978), *Information, Efficiency and the Core of an Economy*, Econometrica, 46, p.807-816

YAMAZAKI, A. (1978), *An Equilibrium Existence Theorem without Convexitiy assumptions*, Econometrica, 46, 541-555

YANNELIS, N.C. (1991), *The Core of an Economy with Differential Information*, Economic Theory, 1, p.183-198

Index

Springer
and the
environment

At Springer we firmly believe that an international science publisher has a special obligation to the environment, and our corporate policies consistently reflect this conviction.

We also expect our business partners – paper mills, printers, packaging manufacturers, etc. – to commit themselves to using materials and production processes that do not harm the environment. The paper in this book is made from low- or no-chlorine pulp and is acid free, in conformance with international standards for paper permanency.

 Springer

Printing: Weihert-Druck GmbH, Darmstadt
Binding: Buchbinderei Schäffer, Grünstadt

Lecture Notes in Economics and Mathematical Systems

For information about Vols. 1–290
please contact your bookseller or Springer-Verlag

Vol. 329: G. Tillmann, Equity, Incentives, and Taxation. VI, 132 pages. 1989.

Vol. 330: P.M. Kort, Optimal Dynamic Investment Policies of a Value Maximizing Firm. VII, 185 pages. 1989.

Vol. 331: A. Lewandowski, A.P. Wierzbicki (Eds.), Aspiration Based Decision Support Systems. X, 400 pages. 1989.

Vol. 332: T.R. Gulledge, Jr., L.A. Litteral (Eds.), Cost Analysis Applications of Economics and Operations Research. Proceedings. VII, 422 pages. 1989.

Vol. 333: N. Dellaert, Production to Order. VII, 158 pages. 1989.

Vol. 334: H.-W. Lorenz, Nonlinear Dynamical Economics and Chaotic Motion. XI, 248 pages. 1989.

Vol. 335: A.G. Lockett, G. Islei (Eds.), Improving Decision Making in Organisations. Proceedings. IX, 606 pages. 1989.

Vol. 336: T. Puu, Nonlinear Economic Dynamics. VII, 119 pages. 1989.

Vol. 337: A. Lewandowski, I. Stanchev (Eds.), Methodology and Software for Interactive Decision Support. VIII, 309 pages. 1989.

Vol. 338: J.K. Ho, R.P. Sundarraj, DECOMP: An Implementation of Dantzig-Wolfe Decomposition for Linear Programming. VI, 206 pages.

Vol. 339: J. Terceiro Lomba, Estimation of Dynamic Econometric Models with Errors in Variables. VIII, 116 pages. 1990.

Vol. 340: T. Vasko, R. Ayres, L. Fontvieille (Eds.), Life Cycles and Long Waves. XIV, 293 pages. 1990.

Vol. 341: G.R. Uhlich, Descriptive Theories of Bargaining. IX, 165 pages. 1990.

Vol. 342: K. Okuguchi, F. Szidarovszky, The Theory of Oligopoly with Multi-Product Firms. V, 167 pages. 1990.

Vol. 343: C. Chiarella, The Elements of a Nonlinear Theory of Economic Dynamics. IX, 149 pages. 1990.

Vol. 344: K. Neumann, Stochastic Project Networks. XI, 237 pages. 1990.

Vol. 345: A. Cambini, E. Castagnoli, L. Martein, P Mazzoleni, S. Schaible (Eds.), Generalized Convexity and Fractional Programming with Economic Applications. Proceedings, 1988. VII, 361 pages. 1990.

Vol. 346: R. von Randow (Ed.), Integer Programming and Related Areas. A Classified Bibliography 1984–1987. XIII, 514 pages. 1990.

Vol. 347: D. Ríos Insua, Sensitivity Analysis in Multiobjective Decision Making. XI, 193 pages. 1990.

Vol. 348: H. Störmer, Binary Functions and their Applications. VIII, 151 pages. 1990.

Vol. 349: G.A. Pfann, Dynamic Modelling of Stochastic Demand for Manufacturing Employment. VI, 158 pages. 1990.

Vol. 350: W.-B. Zhang, Economic Dynamics. X, 232 pages. 1990.

Vol. 351: A. Lewandowski, V. Volkovich (Eds.), Multiobjective Problems of Mathematical Programming. Proceedings, 1988. VII, 315 pages. 1991.

Vol. 352: O. van Hilten, Optimal Firm Behaviour in the Context of Technological Progress and a Business Cycle.

XII, 229 pages. 1991.

Vol. 353: G. Ricci (Ed.), Decision Processes in Economics. Proceedings, 1989. III, 209 pages 1991.

Vol. 354: M. Ivaldi, A Structural Analysis of Expectation Formation. XII, 230 pages. 1991.

Vol. 355: M. Salomon. Deterministic Lotsizing Models for Production Planning. VII, 158 pages. 1991.

Vol. 356: P. Korhonen, A. Lewandowski, J . Wallenius (Eds.), Multiple Criteria Decision Support. Proceedings, 1989. XII, 393 pages. 1991.

Vol. 357: P. Zörnig, Degeneracy Graphs and Simplex Cycling. XV, 194 pages. 1991.

Vol. 358: P. Knottnerus, Linear Models with Correlated Disturbances. VIII, 196 pages. 1991.

Vol. 359: E. de Jong, Exchange Rate Determination and Optimal Economic Policy Under Various Exchange Rate Regimes. VII, 270 pages. 1991.

Vol. 360: P. Stalder, Regime Translations, Spillovers and Buffer Stocks. VI, 193 pages . 1991.

Vol. 361: C. F. Daganzo, Logistics Systems Analysis. X, 321 pages. 1991.

Vol. 362: F. Gehrels, Essays in Macroeconomics of an Open Economy. VII, 183 pages. 1991.

Vol. 363: C. Puppe, Distorted Probabilities and Choice under Risk. VIII, 100 pages . 1991

Vol. 364: B. Horvath, Are Policy Variables Exogenous? XII, 162 pages. 1991.

Vol. 365: G. A. Heuer, U. Leopold-Wildburger. Balanced Silverman Games on General Discrete Sets. V, 140 pages. 1991.

Vol. 366: J. Gruber (Ed.), Econometric Decision Models. Proceedings, 1989. VIII, 636 pages. 1991.

Vol. 367: M. Grauer, D. B. Pressmar (Eds.), Parallel Computing and Mathematical Optimization. Proceedings. V, 208 pages. 1991.

Vol. 368: M. Fedrizzi, J. Kacprzyk, M. Roubens (Eds.), Interactive Fuzzy Optimization. VII, 216 pages. 1991.

Vol. 369: R. Koblo, The Visible Hand. VIII, 131 pages.1991.

Vol. 370: M. J. Beckmann, M. N. Gopalan, R. Subramanian (Eds.), Stochastic Processes and their Applications. Proceedings, 1990. XLI, 292 pages. 1991.

Vol. 371: A. Schmutzler, Flexibility and Adjustment to Information in Sequential Decision Problems. VIII, 198 pages. 1991.

Vol. 372: J. Esteban, The Social Viability of Money. X, 202 pages. 1991.

Vol. 373: A. Billot, Economic Theory of Fuzzy Equilibria. XIII, 164 pages. 1992.

Vol. 374: G. Pflug, U. Dieter (Eds.), Simulation and Optimization. Proceedings, 1990. X, 162 pages. 1992.

Vol. 375: S.-J. Chen, Ch.-L. Hwang, Fuzzy Multiple Attribute Decision Making. XII, 536 pages. 1992.

Vol. 376: K.-H. Jöckel, G. Rothe, W. Sendler (Eds.), Bootstrapping and Related Techniques. Proceedings, 1990. VIII, 247 pages. 1992.

Vol. 377: A. Villar, Operator Theorems with Applications to Distributive Problems and Equilibrium Models. XVI, 160 pages. 1992.

Vol. 378: W. Krabs, J. Zowe (Eds.), Modern Methods of Optimization. Proceedings, 1990. VIII, 348 pages. 1992.

Vol. 379: K. Marti (Ed.), Stochastic Optimization. Proceedings, 1990. VII, 182 pages. 1992.

Vol. 380: J. Odelstad, Invariance and Structural Dependence. XII, 245 pages. 1992.

Vol. 381: C. Giannini, Topics in Structural VAR Econometrics. XI, 131 pages. 1992.

Vol. 382: W. Oettli, D. Pallaschke (Eds.), Advances in Optimization. Proceedings, 1991. X, 527 pages. 1992.

Vol. 383: J. Vartiainen, Capital Accumulation in a Corporatist Economy. VII, 177 pages. 1992.

Vol. 384: A. Martina, Lectures on the Economic Theory of Taxation. XII, 313 pages. 1992.

Vol. 385: J. Gardeazabal, M. Regúlez, The Monetary Model of Exchange Rates and Cointegration. X, 194 pages. 1992.

Vol. 386: M. Desrochers, J.-M. Rousseau (Eds.), Computer-Aided Transit Scheduling. Proceedings, 1990. XIII, 432 pages. 1992.

Vol. 387: W. Gaertner, M. Klemisch-Ahlert, Social Choice and Bargaining Perspectives on Distributive Justice. VIII, 131 pages. 1992.

Vol. 388: D. Bartmann, M. J. Beckmann, Inventory Control. XV, 252 pages. 1992.

Vol. 389: B. Dutta, D. Mookherjee, T. Parthasarathy, T. Raghavan, D. Ray, S. Tijs (Eds.), Game Theory and Economic Applications. Proceedings, 1990. IX, 454 pages. 1992.

Vol. 390: G. Sorger, Minimum Impatience Theorem for Recursive Economic Models. X, 162 pages. 1992.

Vol. 391: C. Keser, Experimental Duopoly Markets with Demand Inertia. X, 150 pages. 1992.

Vol. 392: K. Frauendorfer, Stochastic Two-Stage Programming. VIII, 228 pages. 1992.

Vol. 393: B. Lucke, Price Stabilization on World Agricultural Markets. XI, 274 pages. 1992.

Vol. 394: Y.-J. Lai, C.-L. Hwang, Fuzzy Mathematical Programming. XIII, 301 pages. 1992.

Vol. 395: G. Haag, U. Mueller, K. G. Troitzsch (Eds.), Economic Evolution and Demographic Change. XVI, 409 pages. 1992.

Vol. 396: R. V. V. Vidal (Ed.), Applied Simulated Annealing. VIII, 358 pages. 1992.

Vol. 397: J. Wessels, A. P. Wierzbicki (Eds.), User-Oriented Methodology and Techniques of Decision Analysis and Support. Proceedings, 1991. XII, 295 pages. 1993.

Vol. 398: J.-P. Urbain, Exogeneity in Error Correction Models. XI, 189 pages. 1993.

Vol. 399: F. Gori, L. Geronazzo, M. Galeotti (Eds.), Nonlinear Dynamics in Economics and Social Sciences. Proceedings, 1991. VIII, 367 pages. 1993.

Vol. 400: H. Tanizaki, Nonlinear Filters. XII, 203 pages. 1993.

Vol. 401: K. Mosler, M. Scarsini, Stochastic Orders and Applications. V, 379 pages. 1993.

Vol. 402: A. van den Elzen, Adjustment Processes for Exchange Economies and Noncooperative Games. VII, 146 pages. 1993.

Vol. 403: G. Brennscheidt, Predictive Behavior. VI, 227 pages. 1993.

Vol. 404: Y.-J. Lai, Ch.-L. Hwang, Fuzzy Multiple Objective Decision Making. XIV, 475 pages. 1994.

Vol. 405: S. Komlósi, T. Rapcsák, S. Schaible (Eds.), Generalized Convexity. Proceedings, 1992. VIII, 404 pages. 1994.

Vol. 406: N. M. Hung, N. V. Quyen, Dynamic Timing Decisions Under Uncertainty. X, 194 pages. 1994.

Vol. 407: M. Ooms, Empirical Vector Autoregressive Modeling. XIII, 380 pages. 1994.

Vol. 408: K. Haase, Lotsizing and Scheduling for Production Planning. VIII, 118 pages. 1994.

Vol. 409: A. Sprecher, Resource-Constrained Project Scheduling. XII, 142 pages. 1994.

Vol. 410: R. Winkelmann, Count Data Models. XI, 213 pages. 1994.

Vol. 411: S. Dauzère-Péres, J.-B. Lasserre, An Integrated Approach in Production Planning and Scheduling. XVI, 137 pages. 1994.

Vol. 412: B. Kuon, Two-Person Bargaining Experiments with Incomplete Information. IX, 293 pages. 1994.

Vol. 413: R. Fiorito (Ed.), Inventory, Business Cycles and Monetary Transmission. VI, 287 pages. 1994.

Vol. 414: Y. Crama, A. Oerlemans, F. Spieksma, Production Planning in Automated Manufacturing. X, 210 pages. 1994.

Vol. 415: P. C. Nicola, Imperfect General Equilibrium. XI, 167 pages. 1994.

Vol. 416: H. S. J. Cesar, Control and Game Models of the Greenhouse Effect. XI, 225 pages. 1994.

Vol. 417: B. Ran, D. E. Boyce, Dynamic Urban Transportation Network Models. XV, 391 pages. 1994.

Vol. 418: P. Bogetoft, Non-Cooperative Planning Theory. XI, 309 pages. 1994.

Vol. 419: T. Maruyama, W. Takahashi (Eds.), Nonlinear and Convex Analysis in Economic Theory. VIII, 306 pages. 1995.

Vol. 420: M. Peeters, Time-To-Build. Interrelated Investment and Labour Demand Modelling. With Applications to Six OECD Countries. IX, 204 pages. 1995.

Vol. 421: C. Dang, Triangulations and Simplicial Methods. IX, 196 pages. 1995.

Vol. 422: D. S. Bridges, G. B. Mehta, Representations of Preference Orderings. X, 165 pages. 1995.

Vol. 423: K. Marti, P. Kall (Eds.), Stochastic Programming. Numerical Techniques and Engineering Applications. VIII, 351 pages. 1995.

Vol. 424: G. A. Heuer, U. Leopold-Wildburger, Silverman's Game. X, 283 pages. 1995.

Vol. 425: J. Kohlas, P.-A. Monney, A Mathematical Theory of Hints. XIII, 419 pages, 1995.

Vol. 426: B. Finkenstädt, Nonlinear Dynamics in Economics. IX, 156 pages. 1995.

Vol. 427: F. W. van Tongeren, Microsimulation Modelling of the Corporate Firm. XVII, 275 pages. 1995.

Vol. 428: A. A. Powell, Ch. W. Murphy, Inside a Modern Macroeconometric Model. XVIII, 424 pages. 1995.

Vol. 429: R. Durier, C. Michelot, Recent Developments in Optimization. VIII, 356 pages. 1995.

Vol. 430: J. R. Daduna, I. Branco, J. M. Pinto Paixão (Eds.), Computer-Aided Transit Scheduling. XIV, 374 pages. 1995.

Vol. 431: A. Aulin, Causal and Stochastic Elements in Business Cycles. XI, 116 pages. 1996.

Vol. 432: M. Tamiz (Ed.), Multi-Objective Programming and Goal Programming. VI, 359 pages. 1996.

Vol. 433: J. Menon, Exchange Rates and Prices. XIV, 313 pages. 1996.

Vol. 434: M. W. J. Blok, Dynamic Models of the Firm. VII, 193 pages. 1996.

Vol. 435: L. Chen, Interest Rate Dynamics, Derivatives Pricing, and Risk Management. XII, 149 pages. 1996.

Vol. 436: M. Klemisch-Ahlert, Bargaining in Economic and Ethical Environments. IX, 155 pages. 1996.

Vol. 437: C. Jordan, Batching and Scheduling. IX, 178 pages. 1996.

Vol. 438: A. Villar, General Equilibrium with Increasing Returns. XIII, 164 pages. 1996.

Vol. 439: M. Zenner, Learning to Become Rational. VII, 201 pages. 1996.

Vol. 440: W. Ryll, Litigation and Settlement in a Game with Incomplete Information. VIII, 174 pages. 1996.

Vol. 441: H. Dawid, Adaptive Learning by Genetic Algorithms. IX, 166 pages.1996.

Vol. 442: L. Corchón, Theories of Imperfectly Competitive Markets. XIII, 163 pages. 1996.

Vol. 443: G. Lang, On Overlapping Generations Models with Productive Capital. X, 98 pages. 1996.

Vol. 444: S. Jørgensen, G. Zaccour (Eds.), Dynamic Competitive Analysis in Marketing. X, 285 pages. 1996.

Vol. 445: A. H. Christer, S. Osaki, L. C. Thomas (Eds.), Stochastic Modelling in Innovative Manufacturing. X, 361 pages. 1997.

Vol. 446: G. Dhaene, Encompassing. X, 160 pages. 1997.

Vol. 447: A. Artale, Rings in Auctions. X, 172 pages. 1997.

Vol. 448: G. Fandel, T. Gal (Eds.), Multiple Criteria Decision Making. XII, 678 pages. 1997.

Vol. 449: F. Fang, M. Sanglier (Eds.), Complexity and Self-Organization in Social and Economic Systems. IX, 317 pages, 1997.

Vol. 450: P. M. Pardalos, D. W. Hearn, W. W. Hager, (Eds.), Network Optimization. VIII, 485 pages, 1997.

Vol. 451: M. Salge, Rational Bubbles. Theoretical Basis, Economic Relevance, and Empirical Evidence with a Special Emphasis on the German Stock Market.IX, 265 pages. 1997.

Vol. 452: P. Gritzmann, R. Horst, E. Sachs, R. Tichatschke (Eds.), Recent Advances in Optimization. VIII, 379 pages. 1997.

Vol. 453: A. S. Tangian, J. Gruber (Eds.), Constructing Scalar-Valued Objective Functions. VIII, 298 pages. 1997.

Vol. 454: H.-M. Krolzig, Markov-Switching Vector Autoregressions. XIV, 358 pages. 1997.

Vol. 455: R. Caballero, F. Ruiz, R. E. Steuer (Eds.), Advances in Multiple Objective and Goal Programming. VIII, 391 pages. 1997.

Vol. 456: R. Conte, R. Hegselmann, P. Terna (Eds.), Simulating Social Phenomena. VIII, 536 pages. 1997.

Vol. 457: C. Hsu, Volume and the Nonlinear Dynamics of Stock Returns. VIII, 133 pages. 1998.

Vol. 458: K. Marti, P. Kall (Eds.), Stochastic Programming Methods and Technical Applications. X, 437 pages. 1998.

Vol. 459: H. K. Ryu, D. J. Slottje, Measuring Trends in U.S. Income Inequality. XI, 195 pages. 1998.

Vol. 460: B. Fleischmann, J. A. E. E. van Nunen, M. G. Speranza, P. Stähly, Advances in Distribution Logistic. XI, 535 pages. 1998.

Vol. 461: U. Schmidt, Axiomatic Utility Theory under Risk. XV, 201 pages. 1998.

Vol. 462: L. von Auer, Dynamic Preferences, Choice Mechanisms, and Welfare. XII, 226 pages. 1998.

Vol. 463: G. Abraham-Frois (Ed.), Non-Linear Dynamics and Endogenous Cycles. VI, 204 pages. 1998.

Vol. 464: A. Aulin, The Impact of Science on Economic Growth and its Cycles. IX, 204 pages. 1998.

Vol. 465: T. J. Stewart, R. C. van den Honert (Eds.), Trends in Multicriteria Decision Making. X, 448 pages. 1998.

Vol. 466: A. Sadrieh, The Alternating Double Auction Market. VII, 350 pages. 1998.

Vol. 467: H. Hennig-Schmidt, Bargaining in a Video Experiment. Determinants of Boundedly Rational Behavior. XII, 221 pages. 1999.

Vol. 468: A. Ziegler, A Game Theory Analysis of Options. XIV, 145 pages. 1999.

Vol. 469: M. P. Vogel, Environmental Kuznets Curves. XIII, 197 pages. 1999.

Vol. 470: M. Ammann, Pricing Derivative Credit Risk. XII, 228 pages. 1999.

Vol. 471: N. H. M. Wilson (Ed.), Computer-Aided Transit Scheduling. XI, 444 pages. 1999.

Vol. 472: J.-R. Tyran, Money Illusion and Strategic Complementarity as Causes of Monetary Non-Neutrality. X, 228 pages. 1999.

Vol. 473: S. Helber, Performance Analysis of Flow Lines with Non-Linear Flow of Material. IX, 280 pages. 1999.

Vol. 474: U. Schwalbe, The Core of Economies with Asymmetric Information. IX, 141 pages. 1999.

Vol. 475: L. Kaas, Dynamic Macroelectronics with Imperfect Competition. XI, 155 pages. 1999.